STATISTICAL CALCULATIONS

How To Solve Statistical Problems

E. JOE MIDDLEBROOKS

Dean, College of Engineering
Utah State University
Logan, Utah

ANN ARBOR SCIENCE
PUBLISHERS INC
P.O. BOX 1425 • ANN ARBOR, MICH. 48106

Preface

The detailed mathematical solutions of the statistical analyses presented in this book were developed as a supplemental text for first courses in statistics. Students and practitioners involved in environmental, agricultural and biological sciences and environmental engineering with a minimal background in statistics but with a need to interpret the data gathered during their experiments will benefit most. It frequently becomes difficult to fully understand the concepts and methods presented in statistics unless detailed examples are available for review. The detailed solutions presented in this supplemental text include enough steps to insure that a student will have no difficulty in following the procedures used in the solution to the problem.

I am grateful: to the Literary Executor of the late Sir Ronald A. Fisher, F. R. S., Dr. Frank Yates, F. R. S. and to Longman Group Lts., London, for permission to reprint an abridged Table III from their book *Statistical Tables for Biological, Agricultural and Medical Research* (6th edition, 1974),

to McGraw-Hill Book Company for permission to use the symbols employed by Steel and Torrie: *Principles and Procedures of Statistics,*

to Professor E.S. Pearson and the Biometrika Trustees for permission to use the Tables of F values and the Table of Upper Percentage Points of the Studentized Range presented in the appendix,

to the Iowa State University Press for permission to reproduce the values of the correlation coefficients shown in the appendix,

to The Biometric Society and Dr. D. B. Duncan for permission to use the table of significant studentized ranges for the New Multiple-Range Test.

I am particularly indebted to Dr. W. J. Drapala of Mississippi State University for showing me the value of statistics and for providing the data used in many of the examples.

Special thanks to Kathy Bayn and Barbara South for their careful work in preparing the manuscript.

E. Joe Middlebrooks

To Charlotte and Tracey

Symbols and Meanings[1]

Symbol	Meaning
$\neq$	not equal to; e.g., $3 \neq 4$
$>$	greater than; e.g., $5 > 2$
$\geqslant$	greater than or equal to
$<$	less than; e.g., $3 < 7$
$\leqslant$	less than or equal to
$\| \; \|$	absolute value; e.g., $|-7| = 7$
Σ	sum of
...	indicates a set of obvious missing quantities, e.g., $1,2,...,10$
n!	$n(n-1)...1$ and called n factorial; e.g., $3! = 3(2)\,1 = 6$
$-$	overbar; used to indicate an arithmetic average or mean
$\wedge$	hat; used to indicate an estimate rather than a true value; most often appears over Greek letters
Greek letters	with few exceptions refer to population parameters
μ	population mean
σ^2, σ	population variance and standard deviation
τ, β, etc.	components of population means; commonly used in conjunction with linear models
ϵ	a true experimental error
δ	a true sampling error
β	population regression coefficient
ρ	population correlation coefficient

The preceding Greek letters are also used with subscripts where clarity requires. For example:

$\mu_{\bar{x}}$	mean of a population of $\bar{x}$'s
$\beta_{yx \cdot z}$	regression of Y on X for fixed Z
τ_i	a contribution to the mean of the population receiving the ith treatment

Some exceptions to the use of Greek letters for parameters are:

a	probability of a Type I error

---

[1] From "Principles and Procedures of Statistics" by Robert G. D. Steel and James H. Torrie. Used with permission of McGraw-Hill Book Company.

$1 - a$	confidence coefficient
β	probability of a Type II error
$1 - \beta$	power of a statistical test
χ^2	a common test criterion
Latin letters	used as general symbols, including symbols for sample statistics
X	a variable (Y is also commonly used)
X_i, X_{ij}	individual observations
$X_i, X_{..}$	totals of observations
x_i	$X_i - \bar{x}$
x_i'	$(X_i - \bar{x})/s$
D_j	difference between paired observations
$n, n_{..}$	total sample size
n_{ij}	number of observations in i, jth cell
$\bar{x}, \bar{x}_{..}, \bar{x}_{i.}$	sample means, whole or part of sample
$\bar{\bar{x}}$	mean of a sample of means
$\bar{d}$	$\bar{x}_1 - \bar{x}_2$
$s^2, s_{\bar{x}}^2, s_d^2$	sample variances, unbiased estimators of σ^2, $\sigma_{\bar{x}}^2$, and σ_d^2
$s, s_{\bar{x}}, s_d$	sample standard deviations
$s^2_{y \cdot x}, s_{y \cdot 1 \ldots k}$	sample variances adjusted for regression
CL, CI	confidence limits or interval
l_1, l_2	end points of confidence limits
b	sample regression coefficient
$b_{y1 \cdot 2 \ldots k}$	sample partial regression coefficient
b'	standard regression coefficient
r	sample total or simple correlation coefficient
$r_{12.3 \ldots k}$	sample partial correlation coefficient of X_1 and X_2
$R_{1.2 \ldots k}$	multiple correlation coefficient between X_1 and the other variables
df, f	degrees of freedom
C, CT	correction term
SS	Σx^2_i, sum of squares
MS	mean square
E_a, E_b	error mean squares for split-plot design
E_{yy}, E_{xy}, E_{xx}	error sums of products in covariance (other letters used for other sources of variation)
*	significant, e.g., 2.3*
**	highly significant, e.g., 14.37**
ns	not significant
lsd	least significant difference
RE	relative efficiency
CV	coefficient of variability $(s/\bar{x})100$
$Q = \Sigma c_i T_i$	a comparison where c_i is a constant, T_i is a treatment total, and $\Sigma c_i = 0$
fpc	finite population correction
psu	primary sampling unit

st	stratified, used as subscript
K	Σc_i^2
P	probability
p, 1 - p	probabilities in a binomial distribution
z, t, F	common test criteria
H_0	null hypothesis
H_1	alternate hypothesis, usually a set of alternatives
∞	infinity
C.F.	correction factor for sum of squares calculations
RCB	randomized complete block diagram
CRD	completely random design
P.F.	precision factor

Contents

Introduction

This little book is a collection of detailed solutions to practical statistical problems. No attempt is made to present theory or derivations in the examples of statistical calculations. There are numerous textbooks and popular books on the subject of statistics which provide the theoretical and developmental material. What most of these books are lacking is a detailed presentation of the application of the material presented. This is particularly true in textbooks written for people with good mathematical backgrounds. Frequently examples are abbreviated or include such simple examples that it is impossible to translate this to a more complex analysis encountered in practice. All of the examples presented in this little book are based upon actual experimental results and are easily adaptable to any scientific or engineering field. The examples have been selected from a wide variety of activities in hopes that the reader will see that regardless of the subject area, the application of the statistical technique is identical.

The mathematical symbols employed in this book are the same as those used by Steel and Torrie in their textbook, *Principles and Procedures of Statistics,* published by McGraw-Hill Book Company. These symbols were selected to provide a simple and logical method of mathematical shorthand. The symbols are easily understood and should assist in understanding the examples. For additional information, it is recommended that the above book by Steel and Torrie be consulted. A very interesting and informative statistics book prepared for the general public is entitled *Facts From Figures* by M. J. Moroney published by Penguin Books, Baltimore, Maryland. The book is prepared in paperback form and is inexpensive. The book is written in an informative and humorous manner and provides an excellent insight into the reasons for using statistics.

An attempt has been made to present a detailed example of all of the standard techniques used in analyzing and interpreting scientific and engineering information. The presentations are not exhaustive because of the similarity of various techniques. An example of the similarities is in the Nonlinear Regression and Correlation Calculations. The example includes a log-log or exponential curve fit which plots as a straight line on log-log paper $(Y = aX^b)$. The statistical analysis of all of the log relationships is essentially

1

the same as the linear regression and correlation except that the logarithm is substituted as called for in the linear description of the exponential equation. For example, the exponential equation, $Y = ab^X$, in linear form, log Y = log a + Xlog b, plots as a straight line on the semi-log paper. The regression and correlation are computed by using the logarithms of the experimental values of Y and the unmodified values of X. The procedure spelled out for the application of probability paper is equally applicable to log-probability fits. Use log-probability paper or plot the logarithms versus the probability scale on linear probability paper.

All computer centers have a standard package of programs which will make all of the calculations shown in the examples presented in this book. When this type service is available, it is certainly advisable to use these packaged programs. However, if you wish to know how and in what order these calculations are made, the examples presented in this book will still be of great value to you. Another useful tool that has recently become available is the minicomputer that can be obtained from several firms. Probably the most effective one obtainable is the Hewlett Packard HP-65 which has available a STAT PAC. Practically all of the standard statistical calculations can be made with the HP-65 programmable calculator. In brief, it is advisable to use automatic calculating procedures when possible to avoid errors and to save time. This book is prepared only to demonstrate how the detailed steps are carried out and to help one better understand the analytical procedures.

All of the calculations have been carefully checked to insure that an accurate presentation is made, but in the event that errors were made, it would be appreciated if you would send errata to the author or publisher.

Exercise Number 1

Comparing the Means of Two Materials, Products or Processes by the t-test

Problem:

1. Determine if there is a significant (real) difference between the two treatments using:
   (a) a two-tailed t-test
   (b) a one-tailed t-test
2. Calculate the 95% confidence limits for the true mean difference.
3. For the unpaired data determine whether the variances for each of the treatments are both estimates of the same population variance. Note that if $\sigma_1^2 = \sigma_2^2$ the formula is not identical with that if $\sigma_1^2 \neq \sigma_2^2$ when $n_1 \neq n_2$.

Two Treatments - Not Paired

Two methods were used in a study of the latent heat of fusion of ice. Both Method A (an electrical method) and Method B (a method of mixtures) were conducted with the specimens cooled to -0.72° C. The data represent the change in total heat from -0.72° C to water at 0°C, in calories per gram of mass.

Method A   79.98, 80.04, 80.02, 80.04, 80.03, 80.03, 80.04, 79.97, 80.05, 80.03, 80.02, 80.00, 80.02.

Method B   80.02, 79.94, 79.98, 79.97, 80.03, 79.95, 79.97, 79.97.

Two Treatments - Paired

The following results were reported by Hirschboeck (Proc. Soc. Expt. Biol. Med., 47:311.1941) of the clotting times observed in 10 pairs of blood samples. One out of each pair was chosen at random and treated with methacrylate and the other was treated with paraffin.

| | Clotting Time in Minutes | | |
Sample	Paraffin	Methacrylate	Difference
1	10	13	-3
2	27	20	7
3	11	9	2
4	18	12	6
5	19	11	8
6	16	14	2
7	16	19	-3
8	18	12	6
9	22	11	11
10	26	18	8

Purpose:

To determine if there is a significant difference between two methods of determining the latent heat of fusion of ice in which the observations are not paired, and to determine if there is a significant difference in the clotting times produced by paraffin and methacrylate with the observations paired.

Method and Materials:

A two-tailed t-test and a one-tailed t-test were used to compare the two treatments employed in both experiments. Data from the latent heat of fusion of ice experiment represent the change in total heat from $-0.72°$ C to water at $0°$C in calories per gram of mass. The clotting experimental data represent clotting times in 10 pairs of blood samples. One blood sample out of each pair was chosen at random and treated with methacrylate and the other was treated with paraffin.

Results and Computations:

Data:

Table 1-1. Latent Heat of Fusion of Ice, Unpaired Data

"A" Electrical Method calories/gram of mass	"B" Method of Mixtures calories/gram of mass
79.98	80.02
80.04	79.94
80.02	79.98
80.04	79.97
80.03	80.03
80.03	79.95
80.04	79.97
79.97	79.97
80.05	
80.03	
80.02	
80.00	
80.02	

Σ_{Aj} = 1040.27 Σ_{Bj} = 639.83

$\bar{x}_A$ = 80.0208 $\bar{x}_B$ = 79.9787

Calculation of the Standard Error of the Differences, $\bar{d}$:

$$SS_A = \sum_{j}^{13} X_{Aj}^2 - (\sum_{j}^{13} X_{Aj})^2/13 = \sum_{j}^{13} x_{Aj}^2$$

$$SS_A = (79.98^2 + ... + 80.02^2) - (1040.27)^2/13$$

$$SS_A = 83,243.2125 - 1,082,161.6729/13$$

$$SS_A = 83,243.2125 - 83,243.2056$$

$$SS_A = \underline{0.0069}$$

$$s_A^2 = SS_A/n_A-1 = 0.0069/12 = \underline{0.000575}$$

$$SS_B = \sum_{j}^{8} X_{Bj}^2 - (\sum_{j}^{8} X_{Bj})^2/8 = \sum_{j}^{8} x_{Bj}^2$$

$$SS_B = (79.94^2 + \ldots + 79.97^2) - (639.83)^2/8$$

$$SS_B = 51{,}172.8105 - 409{,}382.4289/8$$

$$SS_B = 51{,}172.8105 - 51{,}172.8036$$

$$SS_B = \underline{0.0069}$$

$$s_B^2 = SS_B/n_B\text{-}1 = 0.0069/7 = \underline{0.000986}$$

$$F = \frac{s_B^2}{s_A^2} = \frac{0.000986}{0.000575} = \underline{\underline{1.71}}$$

$$F_{.025}\,(7,12) = 3.61 \qquad \text{(Obtained from Table A-2)}$$

Since the calculated F-value is smaller than the critical F-value, there is no significant difference in the variances; therefore, the variances can be pooled.

$$s^2 = \frac{\Sigma\, x_{Aj}^2 + \Sigma\, x_{Bj}^2}{n_1 - 1 + n_2 - 1} = \frac{0.0069 + 0.0069}{12 + 7}$$

$$s^2 = \underline{0.0007263}$$

$$s_{\bar{d}} = \sqrt{s^2 \left\{ \frac{n_1 + n_2}{n_1 n_2} \right\}} = \sqrt{0.0007263 \left\{ \frac{13+8}{13(8)} \right\}}$$

$$s_{\bar{d}} = \sqrt{0.00014666} = \underline{0.0121}$$

Two-tailed t-test:

$$t = \frac{(\bar{x}_A - \mu_A) - (\bar{x}_B - \mu_B)}{s_{\bar{d}}}$$

$$H_0 : \mu_1 = \mu_2 \ \therefore$$
$$H_1 : \mu_1 \neq \mu_2$$

$$t = \frac{\bar{x}_A - \bar{x}_B}{s_{\bar{d}}} = \frac{\bar{d}}{s_{\bar{d}}}$$

$$t = \frac{80.0208 - 79.9787}{0.0121} = \underline{\underline{3.471}}**$$

$$t_{.05} \left\{ 19df \right\}_{.01} \begin{matrix} = 2.093 \\ \\ = 2.861 \end{matrix} \qquad \text{(Obtained from Table A-1)}$$

The experimental value of t lies outside the critical tabular value; therefore, the $H_0 : \mu_1 = \mu_2$ is rejected and it is claimed that $\bar{x}_A$ differs significantly from $\bar{x}_B$.

95% Confidence Limits for the True Mean Difference:

$$\bar{d} \pm t_{.05} \ (19 d.f.) \times s_{\bar{d}}$$

$$0.042 \pm 2.093 \ (0.0121)$$

$$\underline{\underline{0.042 \pm 0.025}}$$

One-tailed t-test:

$$H_0 : \mu_A \leqslant \mu_B$$

$$H_1 : \mu_A > \mu_B$$

$$t_{.10} \ (19df) = 1.729 \qquad \text{(Obtained from Table A-1)}$$

$$t_{.02} \ (19df) = 2.539$$

$$t = 3.471**$$

The experimental value of t lies outside the critical tabular value; therefore, the $H_1 : \mu_A > \mu_B$ is accepted and H_0 is rejected.

95% Confidence Limits for the True Mean Difference:

$$\bar{d} \pm t_{.10} \ (19 d.f.) \times s_{\bar{d}}$$

$$0.042 \pm 1.729 \ (0.0121)$$

$$\underline{\underline{0.042 \pm 0.021}}$$

Data:

Table 1-2. Clotting Times

Sample	Paired Data		Difference	Difference^2
	Paraffin	Methacrylate		
1	10	13	-3	9
2	27	20	7	49
3	11	9	2	4
4	18	12	6	36
5	19	11	8	64
6	16	14	2	4
7	16	19	-3	9
8	18	12	6	36
9	22	11	11	121
10	26	18	8	64
		Σ	44	396

$$\bar{d} = 4.40$$

Calculation of the Standard Error of the Differences:

$$SSD = \sum_{j}^{10} D_j^2 - (\Sigma D_j)^2 / 10$$

$$SSD = (-3^2 + ... + 8^2) - (44)^2 / 10$$

$$SSD = 396 - 193.6$$

$$SSD = \underline{202.4}$$

$$s_{\bar{d}} = \sqrt{\frac{SSD}{n(n-1)}}, \text{ where } n = \text{no. of pairs}$$

$$s_{\bar{d}} = \sqrt{\frac{202.4}{10(10-1)}} = \sqrt{2.25} = \underline{\underline{1.50}}$$

Two-tailed t-test:

$$t = \frac{\bar{d}}{s_{\bar{d}}} = \frac{4.40}{1.50} = \underline{\underline{2.933*}}$$

$H_0 : \mu_p = \mu_m$

$$t_{.05} \ (9 \text{d.f.}) = 2.262$$
$$t_{.01} \ (9 \text{d.f.}) = 3.250 \qquad \text{(Obtained from Table A-1)}$$

The calculated value of t is larger than the critical tabular value at the 95% level; therefore, $\mu_1 \neq \mu_2$, or there is a significant difference in the two means.

95% Confidence Limits For the True Mean Difference:

$\bar{d} \pm t_{.05} \ (9\text{d.f.}) \times s_{\bar{d}}$

$4.40 \pm 2.262 \ (1.50)$

$\underline{4.40 \pm 3.39}$

One-tailed t-test:

$H_1 : \mu_p > \mu_m$

$$t_{.10} \ (19\text{d.f.}) = 1.729$$
$$t_{.02} \ (19\text{d.f.}) = 2.539 \qquad \text{(Obtained from Table A-1)}$$

$t = 2.933*$

The calculated t-value is greater than the critical tabular value; therefore, the H_1 is accepted, $\mu_p > \mu_m$.

95% Confidence Limits for the True Mean Difference:

$\bar{d} \pm t_{0.10} \ (9\text{d.f.}) \times s_{\bar{d}}$

$4.40 \pm 1.729 \ (1.50)$

$\underline{4.40 \pm 2.59}$

Conclusions:

Real differences exist between the two methods of measuring the heat of fusion of ice and the electrical method will give the larger value for the mean heat of fusion. In the paired experiment a significant difference exists between the two methods of clotting blood.

Exercise Number 2

Inferences Concerning More Than Two Means Analysis of Variance: Completely Random Design

Problem:

1. Indicate the nature of the Null Hypothesis (H_o) used and any assumptions underlying the variances.
2. Analyze the data by means of the analysis of variance technique.
3. Demonstrate the additive nature of the sum of squares by determining the error sum of squares directly.
4. Calculate the coefficient of variability (variation).
5. Even if the treatment F-value is non-significant, calculate the least significant difference at the 5% level of probability and illustrate its use. In these problems it will be necessary to select arbitrarily one of the treatments as a standard or check.
6. Make a creditable interpretation of the results.

The 24-hour water absorption (in percent of dry weight) of samples of concrete taken from five different types of precast concrete curbs made with different aggregates is shown in the table below:

Curb Type					Total	Mean
A	6.7	5.8	5.8	5.5		
B	5.1	4.7	5.1	5.2		
C	4.4	4.9	4.6	4.5		
D	6.7	7.2	6.8	6.3		
E	6.5	5.8	4.7	5.9		

Purpose:

To determine if there is any effect on the water absorption characteristics of concrete which was made with different aggregates, and to evaluate the virulence or toxicity of three typhoid organisms, 9D, 11C and DSCl.

Methods and Materials:

A completely random design was used to evaluate the data. The data for the concrete curb experiments were recorded as 24-hour water absorption in percent of dry weight, and the virulence data were recorded as days till death in mice injected with the three types of typhoid organisms.

Results and Computations:

Data:

Table 2-1. Concrete Curb Water Absorption

Curb Type	Replications of 24 hr Water Absorption %				Total	Mean	$\sum\limits_{j}^{4} X_{ij}^2$
A	6.7	5.8	5.8	5.5	23.8	5.950*	142.42
B	5.1	4.7	5.1	5.2	20.1	5.025ⓢ	101.15
C	4.4	4.9	4.6	4.5	18.4	4.600	84.78
D	6.7	7.2	6.8	6.3	27.0	6.750**	182.66
E	6.5	5.8	4.7	5.9	22.9	5.725*	132.79
Total					112.2		643.80

$\bar{x}_{..} = \underline{\underline{5.61}}$

Null Hypothesis:

$$H_0 : \mu_1 = \mu_2 = \mu_3 = \mu_4 = \mu_5$$

Analysis of Variance Calculations:

Total Sum of Squares:

$$TSS = \sum\limits_{i}^{5} \sum\limits_{j}^{4} X_{ij}^2 - X_{..}^2/20$$

$$TSS = (6.7^2 + ... + 5.9^2) - (112.2)^2/20$$

$$TSS = 643.80 - 629.44 = \underline{\underline{14.36}}$$

Curb Sum of Squares:

$$CSS = \sum_{i}^{5} X_{i.}{}^{2}/4 - C.\ F.$$

$$CSS = \frac{(23.8)^2 + (20.1)^2 + (18.4)^2 + (27.0)^2 + (22.9)^2}{4} - \frac{(112.2)^2}{20}$$

$$CSS = 640.60 - 629.44 = \underline{\underline{11.16}}$$

Error Sum of Squares:

$$ESS = TSS - CSS$$

$$ESS = 14.36 - 11.16 = \underline{\underline{3.20}}$$

Mean Square:

$$MS_C = \frac{CSS}{d.f.} = \frac{11.16}{5 - 1} = \underline{\underline{2.790}}$$

$$MS_E = \frac{ESS}{d.f.} = \frac{3.20}{16 - 1} = \underline{\underline{0.213}}$$

Calculated F-value:

$$F = \frac{MS_C}{MS_E} = \frac{2.79}{0.21} = \underline{\underline{13.29**}}$$

Critical F-value From Table A-2:

$$F_{.05} = 3.06$$
$$F_{.01} = 4.89$$

There is a highly significant difference between the curb type means; therefore, the hypothesis is rejected.

Table 2-2. Analysis of Variance Summary

Source	d.f.	SS	MS	F
Curbs	4	11.16	2.79	13.29**
Error	15	3.20	0.21	
Total	19	14.36		

Error Sum of Squares Computed Individually To Demonstrate the Additive Nature of the Sum of Squares:

Curb A

$$SS = \sum_{j}^{4} X_{1j}^{2} - X_{1.}^{2}/4$$

$$SS = 142.42 - (23.8)^{2}/4$$

$$SS = 142.42 - 141.61 = \underline{0.81}$$

Curb B

$$SS = \sum_{j}^{4} X_{2j}^{2} - X_{2.}^{2}/4$$

$$SS = 101.15 - (20.1)^{2}/4$$

$$SS = 101.15 - 101.02 = \underline{0.13}$$

Curb C

$$SS = 84.78 - (18.4)^{2}/4$$

$$SS = 84.78 - 84.64 = \underline{0.14}$$

Curb D

$$SS = 182.66 - (27.0)^{2}/4$$

$$SS = 182.66 - 182.25 = \underline{0.41}$$

Curb E

$$SS = 132.79 - (22.9)^{2}/4$$

$$SS = 132.79 - 131.10 = \underline{1.69}$$

$$\text{Pooled Error SS} = \sum_{i}^{t} (\sum_{j}^{r} X_{ij}^{2} - X_{i.}^{2}/r)$$

$$\text{Error SS} = 0.81 + 0.13 + 0.14 + 0.41 + 1.69$$

$$\text{Error SS} = 3.18$$

Difference of 0.02 is due to rounding error.

Coefficient of Variability:

$$C.V. = \frac{\sqrt{s^2}}{\bar{x}_{..}} \times 100$$

Where s^2 is the error mean square.

$$C.V. = \frac{\sqrt{0.21}}{5.61} \times 100 = \underline{8.2\%}$$

Least Significant Difference:

$$LSD_{.05 \atop .01} = t_{.05 \atop .01} \{t(r-1)\} \times s_{\bar{d}} \qquad \text{(t-values obtained from Table A-1)}$$

$$s_{\bar{d}} = \sqrt{\frac{2s^2}{r}} = \sqrt{\frac{2(0.21)}{4}} = \sqrt{0.105} = \underline{\underline{0.324}}$$

$$LSD_{.05} = 2.131 \times 0.324 = \underline{\underline{0.690}}$$

$$LSD_{.01} = 2.947 \times 0.324 = \underline{\underline{0.955}}$$

Curb B was selected as a standard with which to compare the other means. The comparisons are shown in Table 2-1 with the following notation being used:

  Significant difference - *
  Highly Significant difference - **
  Standard -   Ⓢ

Virulence Experiment

Data:

Table 2-3. Days Till Death of Mice Inoculated With 9D, 11C and DSCl Typhoid Organisms

Days Till Death	Number of Mice Inoculated With Indicated Strain			Total
	9D	11C	DSCl	
2	6	1	3	10
3	4	3	5	12
4	9	3	5	17
5	8	6	8	22
6	3	6	19	28
7	1	14	23	38
8		11	22	33
9		4	14	18
10		6	14	20
11		2	7	9
12		3	8	11
13		1	4	5
14			1	1
Total mice/ strain	31	60	133	224
ΣX	125	442	1037	1604
ΣX^2	561	3602	8961	13124
$\bar{x}$	4.03	7.37	7.80	7.16

Null Hypothesis:

$$H_0 : \mu_1 = \mu_2 = \mu_3$$

Analysis of Variance Calculations:

Total Sum of Squares:

$$TSS = \sum_i^t \sum_j^r (X_{ij}^2 - X_{..}^2 / \sum_i^t r_i)$$

$$TSS = 13,124 - (1604)^2 / 224$$

$$TSS = 13,124 - 11,485.785 = \underline{1,638.215}$$

Organisms Sum of Squares:

$$OSS = \sum_i^t X_{i.}^2 / r_i - C.F.$$

$$OSS = \left[\frac{(125)^2}{31} + \frac{(442)^2}{60} + \frac{(1037)^2}{133} \right] - 11,485.785$$

$$OSS = \underline{359.79}$$

Error Sum of Squares:

$$ESS = TSS - OSS$$

$$ESS = 1638.215 - 359.79 = \underline{1,278.43}$$

Mean Square:

$$MS_O = \frac{OSS}{d.f.} = \frac{359.79}{2} = \underline{179.895}$$

$$MS_E = \frac{ESS}{d.f.} = \frac{1,278.43}{221} = \underline{5.785}$$

Calculated F-value:

$$F = \frac{MS_O}{MS_E} = \frac{179.895}{5.785} = \underline{31.098}**$$

Critical F-value From Table A-2:

$$F_{.05} = 3.00$$

$$F_{.01} = 4.61$$

The calculated F-value exceeds the critical $F_{.01}$ value which indicates there is a highly significant difference between the means, or the hypothesis is rejected.

Table 2-4. Analysis of Variance Summary

Source	d.f.	SS	MS	F
Organisms	2	359.79	179.895	31.10**
Error	221	1,278.43	5.785	
Total	223	1,638.22		

Error Sum of Squares Computed Individually to Demonstrate The Additive Nature of the Sum of Squares:

Strain 9D

$$SS = \sum_{j}^{31} X_{1j}^2 - X_{1.}^2/31$$

$$SS = 561 - (125)^2/31$$

$$SS = 561 - 504.03 = \underline{56.97}$$

Strain 11C

$$SS = \sum_{j}^{60} X_{2j}^2 - X_{2.}^2/60$$

$$SS = 3602 - (442)^2/60 = \underline{345.93}$$

Strain DSCl

$$SS = \sum_{j}^{133} X_{3j}^2 - X_{3.}^2/133$$

$$SS = 8961 - (1037)^2/133$$

$$SS = 8961 - 8085.48 = \underline{875.52}$$

Pooled Error S.S. $= \sum_{i}^{t} (\sum_{j}^{r} X_{ij}^{2} - X_{i.}^{2}/r)$

Error Sum of Squares $= 56.97 + 345.93 + 875.52$

$$ESS = \underline{1278.42}$$

Coefficient of Variability:

$$C.V. = \frac{\sqrt{s^2}}{\bar{x}} \times 100$$

$$C.V. = \frac{\sqrt{5.785}}{7.16} \times 100 = \frac{2.405 \times 100}{7.16} = \underline{\underline{33.6\%}}$$

Least Significant Difference:

$\bar{x}_{11C}$ was selected as a standard.

$\bar{x}_{11C}$ vs $\bar{x}_{9D}$

$$L.S.D._{\substack{.05 \\ .01}} = t_{\substack{.05 \\ .01}} \{t(r-1)\} \times s_{\bar{d}}$$

$$s_{\bar{d}} = \sqrt{\frac{s^2}{31} + \frac{s^2}{60}} = \sqrt{\frac{5.785}{31} + \frac{5.785}{60}}$$

$$s_{\bar{d}} = \sqrt{.18661 + .09642} = \sqrt{0.28303} = \underline{\underline{0.532}}$$

Tabular Value of t:

$t_{.05} = 1.960$

(Table A-1)

$t_{.01} = 2.576$

$LSD_{.05} = 1.960 \times 0.532 = 1.043$

$$LSD_{.01} = 2.576 \times 0.532 = 1.3704$$

$$\bar{x}_{11C} - \bar{x}_{9D} = 7.37 - 4.03 = \underline{\underline{3.34}}$$

The calculated difference exceeds the $LSD_{.01}$ value; therefore, a highly significant difference exists between the strains 11C and 9D.

$\bar{x}_{11C}$ vs $\bar{x}_{DSCl}$

$$s_{\bar{d}} = \sqrt{\frac{5.785}{133} + \frac{5.785}{60}} = \sqrt{.04350 + .09642}$$

$$s_{\bar{d}} = \sqrt{.13992} = \underline{\underline{0.3741}}$$

$$LSD_{.05} = 1.960 \times .3741 = 0.7332$$

$$LSD_{.01} = 2.576 \times .3741 = 0.9637$$

$$\bar{x}_{DSCl} - \bar{x}_{11C} = 7.80 - 7.37 = \underline{\underline{0.43}}$$

The calculated difference does not exceed the $LSD_{.05}$; therefore, there is no significant difference between the effects of strains 11C and DSC1.

Conclusions:

Real differences exist between the water adsorption qualities of all the curbs except the curbs B and C with curb B selected as the standard for comparison. Real differences exist among the virulence of the three strains of organisms as far as their ability to induce death in mice was concerned. The typhoid strain 9D differed significantly from 11C - the mean response for 9D was 4.03 days till death and that for 11C was 7.37 days till death.

Exercise Number 3

The Separation of Treatment Means Using Duncan's New Multiple Range Procedure

Problem 1. Equal Number of Replications:

A CRD was used to evaluate the amount of fat absorbed in grams by doughnuts cooked in 8 different kinds of fat. Six doughnuts were tested in each fat.

The ANOV table gave:

Source	df	MS
Fats	7	503.9
Residual	40	141.6
Total	47	

The fat means ranked in order are:

G	H	E	A	F	B	C	D
161	162	165	172	176	178	182	185

Problem 2. Unequal Number of Replications:

Five treatments were evaluated using a CRD where the results were as follows:

Source	df	MS
Treatments	4	20.10
Error	59	3.36

Treatment means ranked in order and number of specimens per treatment:

	1	4	2	3	5
x̄	63.2	63.5	63.6	64.2	66.1
reps	11	24	8	6	15

Determine the significant differences among the means at P = .05 in the foregoing two examples.

References:

1.   Duncan, D. B. Multiple Range and Multiple F Tests. Biometrics 11:1-42. 1955.

2.   Kramer, C. Y. Extension of Multiple Range Tests To Group Means with Unequal Numbers of Replications. Biometrics 12:307-310. 1956.

Purpose:

To determine if significant differences exist between the means of treatments with one experiment performed with an equal number of replications and the other containing an unequal number of replications.

Methods and Materials:

A completely random design was used to evaluate the data, and the means were compared with the Duncan's New Multiple Range Procedure. The data from the first experiment were recorded as the amount of fat absorbed in grams by doughnuts cooked in 8 different kinds of fats. The data were not identified in the second experiment.

Results and Computations:

Experiment 1. Equal Number of Replications

Table 3-1. The Analysis of Variance Summary

Source	d.f.	M.S.
Fats	7	503.9
Error	40	141.6
Total	47	

Table 3-2. Ranked Means

Mean	G	H	E	A	F	B	C	D
	161	162	165	172	176	178	182	185

Standard Error of the Means:

$$s_{\bar{x}} = \sqrt{\frac{s^2}{r}} = \sqrt{\frac{141.6}{6}} = \sqrt{23.60} = \underline{\underline{4.86}}$$

p   =   number of means in a comparison

$SSR_{.05}$ =   significant studentized range for the 5% level. Obtained from Table A-3 in the appendix.

LSR   =   least significant range

LSR   = $s_{\bar{x}}$ x $SR_{.05}$

Table 3-3. Least Significant Range Values at the 5% Level

Value of p	2	3	4	5	6	7	8
$SSR_{.05}$	2.86	3.01	3.10	3.17	3.22	3.27	3.30
LSR	14	15	15	15	16	16	16

The least significant range for p = 8 is subtracted from the largest mean, 185 - 16 = 169. All means less than 169 are then declared significantly different from the largest since the LSR decreases with the number of means in the range being tested. The testing procedure is continued with the mean adjacent to the largest until all are tested. The results are summarized in Table 3-2 by drawing a line under the means that do not differ significantly. **Means not connected by underscoring with the same line are significantly different.**

Experiment 2. Unequal Number of Replications

Table 3-4. The Analysis of Variance Summary

Source	d.f.	MS
Treatments	4	20.10
Error	59	3.36

Table 3-5. Ranked Means and Number of Specimens Per Treatment

Treatment	1	4	2	3	5
$\bar{x}$	63.2	63.5	63.6	64.2	66.1
reps	11	24	8	6	15

When $n_1 \neq n_2 \neq n_3 \neq n_4$, etc.

$$s_{\bar{x}} = \frac{1}{\sqrt{2}} \sqrt{\frac{n_1 + n_2}{n_1 n_2} s^2}$$

$$LSR_{.05} = s_{\bar{x}} \times SR \text{ for } Z_p, n_2$$

Suppose that five means are being compared. In order that the difference between $\bar{x}_5 - \bar{x}_1$ be real, this difference must exceed or be equal to

$$\sqrt{\frac{n_1 + n_5}{2(n_1 n_5)}} s^2 \cdot Z_{5}, n_2 \quad \text{where } n_2 \text{ equals the degrees of freedom.}$$

i.e. $\bar{x}_5 - \bar{x}_1 \geqq \sqrt{\dfrac{n_1 + n_5}{2(n_1 n_5)}} \; s \cdot Z_{5}, n_2$

This can be rewritten as

$$\bar{x}_5 - \bar{x}_1 \sqrt{\frac{2(n_1 n_5)}{n_1 + n_5}} \geqq s \cdot Z_{5}, n_2$$

$$LSR = Rp = s_{\bar{x}} \times S.R. \text{ (equal numbers)}$$

$$Rp' = s \times S.R. \text{ (unequal numbers)}$$

The SR's are taken from Table A-3 of the appendix as before and the Rp' table is completed. It is necessary to make individual calculations for each comparison when unequal numbers are used. Significant differences are denoted by an asterisk and non-significance is shown by the letters n.s.

$$s = \sqrt{3.36} = 1.833 = \underline{\underline{1.83}}$$

Table 3-6. Least Significant Range Values at the 5% Level

Value of p	2	3	4	5
$n_2 = 59$	2.83	2.98	3.08	3.14
Rp'	5.18	5.45	5.64	5.75

Comparison of Means:

$\bar{x}_5$ vs $\bar{x}_1$

In order that $\bar{x}_5 - \bar{x}_1$ be significant

$$\bar{x}_5 - \bar{x}_1 \sqrt{\frac{2(n_5 n_1)}{n_5 + n_1}} \geqslant R_5'$$

$$(66.1 - 63.2)\sqrt{\frac{2(15)(11)}{11 + 15}} = 2.9\sqrt{12.692} = 2.9(3.562)$$

$$= \underline{\underline{10.3^* > R_5'}}$$

$\bar{x}_5$ vs $\bar{x}_4$

$$(66.1 - 63.5)\sqrt{\frac{2(15)(24)}{15 + 24}} = 2.6\sqrt{18.462} = 2.6(4.297)$$

$$= \underline{\underline{11.2^* > R_4'}}$$

$\bar{x}_5$ vs $\bar{x}_2$

$$(66.1 - 63.6)\sqrt{\frac{2(15)(8)}{15 + 8}} = 2.5\sqrt{10.435} = 2.5(3.229)$$

$$= 8.1^* > R'_3$$

$\bar{x}_5$ vs $\bar{x}_3$

$$(66.1 - 64.2)\sqrt{\frac{2(15)(6)}{15 + 6}} = 1.9\sqrt{8.571} = 1.9(2.928)$$

$$= 5.56^* > R'_2$$

$\bar{x}_3$ vs $\bar{x}_1$

$$(64.2 - 63.2)\sqrt{\frac{2(6)(11)}{6 + 11}} = 1.0\sqrt{7.706} = 1.0(2.775)$$

$$= 2.78^{n.s.} < R'_4$$

Obviously if $\bar{x}_3$ does not differ significantly from $\bar{x}_1$, it does not differ from means larger than $\bar{x}_1$.

$\bar{x}_2$ vs $\bar{x}_1$

$$(63.6 - 63.2)\sqrt{\frac{2(8)(11)}{8 + 11}} = 0.4\sqrt{9.264} = 0.4(3.044)$$

$$= 1.22^{n.s.} < R'_3$$

Therefore, $\bar{x}_2$ does not differ from means larger than $\bar{x}_1$.

Table 3-7. Ranked Means Comparisons

Mean	1	4	2	3	5
$\bar{x}$	63.2	63.5	63.6	64.2	66.1

Conclusions:

Significant differences exist among the fat absorption qualities of the dough-nuts cooked in different kinds of fats. Differences exist between treatment 5 and the other means of the experiment 2, but the others are the same, or do not exhibit significant differences.

Exercise Number 4

Randomized Complete Block Design (RCB)

Problem:

Hutson, R. Tests of Lindane and other insecticides for control of *Lygus oblineatus*. Jour. Econ. Ent. 44:773-779. 1955.

Insecticide		Replication Number						Totals	Means
		1	2	3	4	5	6		
Control		62	55	50	45	52	49		
Parathion	15%	22	30	25	29	28	18		
Lindane	25%	13	17	11	24	22	14		
TEPP	50%	26	23	20	27	25	34		
Totals									

The above data are the number of punctures in the midrib of Chinese cabbage produced by *L. oblineatus*.

Determine the following:
1. Analyze one example only using the ANOV procedure.
2. Use Tukey's w-procedure to determine significant differences among treatment means.
3. Write out the expectations of the mean squares using the mixed model where the treatments are fixed and the replications are random.
4. Calculate the C.V.
5. Compare the precision of the randomized complete block design to that expected if a completely random design had been used.

In addition to the foregoing answer determine the following, given the following ANOV table.

Source of Variation	df	Mean Square	Expected Mean Square
Replications	4	240	
Treatments	5	360	
Experimental error	20	120	
Sampling error	150	30	
Determinations: samples	180	4	

1. Determine the number of treatments, replications, samples within plots and the number of determinations used in each sample within each replication within each treatment.
2. Give the numerical value for the experimental error mean square in the above analysis for the following:
   a. if 10 had been added to each determination.
   b. if each determination had been multiplied by the constant 10.
3. Fill in the expected mean squares in the above table, assuming you are interested in just these 6 treatments but that replications, samples and determinations may be considered as random variables.
4. Compute the variance of a treatment mean and show how this statistic can be employed in Tukey's w-procedure.
5. Estimate the gain or loss in efficiency in the above experiment if you had taken 8 samples per plot and had made only 1 determination per sample.

Purpose:

To determine the effectiveness of several insecticides for controlling *Lygus oblineatus.* Also it was necessary to provide various factors for data that had been reduced and presented in an ANOV table.

Method and Materials:

A randomized complete block design was used to evaluate the data. The data are recorded as the number of punctures in the midrib of Chinese cabbage produced by *Lygus oblineatus.* The data for the second problem were not defined.

Results and Computations:

Data:

Table 4-1. Number Punctures in Midrib of Chinese Cabbage by *L. oblineatus*

Insecticide		\multicolumn{6}{c}{Replication Number}						Totals	Means
		1	2	3	4	5	6		
Control		62	55	50	45	52	49	313	52.167
Parathion	15%	22	30	25	29	28	18	152	25.333
Lindane	25%	13	17	11	24	22	14	101	16.805
TEPP	50%	26	23	20	27	25	34	155	25.833
Totals		123	125	106	125	127	115	721	30.042

Analysis of Variance Calculations:

Total Sum of Squares:

$$TSS = \sum_i^t \sum_j^r X_{ij}^2 - X_{..}^2/tr$$

$$TSS = (62^2 + \dots + 34^2) - (721)^2/24$$

$$TSS = 26{,}407 - 21{,}660.04$$

$$TSS = \underline{4{,}746.96}$$

Treatment Sum of Squares:

$$TrSS = \sum_i^t X_{i.}^2/r - C.F.$$

$$TrSS = \frac{(313^2 + 152^2 + 101^2 + 155^2)}{6} - 21{,}660.04$$

$$TrSS = \frac{155{,}299}{6} - 21{,}660.04$$

$$TrSS = 25{,}883.16 - 21{,}660.04$$

$$TrSS = \underline{4{,}223.12}$$

Replication Sum of Squares:

$$RSS = \sum_j^r X_{.j}^2/t - C.F.$$

$$RSS = \frac{(123^2 + \dots + 115^2)}{4} - 21{,}660.04$$

$$RSS = \frac{86,969}{4} - 21,660.04 = 21,742.25 - 21,660.04$$

$$RSS = \underline{82.21}$$

Error Sum of Squares:

$$ESS = TSS - (TrSS + RSS)$$

$$ESS = 4,746.96 - (4,223.12 + 82.21)$$

$$ESS = \underline{441.63}$$

Mean Squares:

$$MS_T = \frac{TrSS}{d.f.} = \frac{4223.12}{3} = \underline{1,407.707}$$

$$MS_R = \frac{RSS}{d.f.} = \frac{441.63}{5} = \underline{88.326}$$

$$MS_E = \frac{ESS}{d.f.} = \frac{82.21}{15} = \underline{5.481}$$

Calculated F-value:

$$F = \frac{MS_T}{MS_E} = \frac{1,407.707}{29.442} = \underline{47.813**}$$

Critical F-value From Table A-2:

$$F_{(3,15).05} = 3.29$$

$$F_{(3,15).01} = 5.42$$

The calculated F-value exceeds the critical value at the 1% level; therefore, there exists a highly significant difference between the treatments. It is assumed or known there exists a difference between the replications.

Table 4-2. Analysis of Variance Summary

Source	d.f.	S.S.	MS	F
Treats.	3	4,223.12	1,407.707	47.813**
Reps.	5	82.21	16.442	
Error	15	441.63	29.442	
Total	23	4,746.96		

Standard Error of the Means:

$$s_{\bar{x}} = \sqrt{\frac{s^2}{r}} \quad \text{where } s^2 = \text{Error Mean Square}$$

$$s_{\bar{x}} = \sqrt{\frac{29.442}{6}} = \sqrt{4.9070} = \underline{2.2151}$$

Comparison of Means by Tukey's w-procedure

w = h.s.d.$_{.05}$ = $s_{\bar{x}}$ x q$_{.05}$: p, n$_2$ (q obtained from Table A-4)
p = number of treatments = 4
n$_2$ = error degrees of freedom = 15
w = 2.215 x 4.08 = 9.037

Table 4-3. Ranked Means

Insecticide	Lindane 25%	Parathion 15%	TEPP 50%	Control
Mean	16.805	25.333	25.833	52.167

52.167 - 9.037 = 43.130   25.333 - 9.037 = 16.796

Means exhibiting no significant differences are shown in Table 4-3 by drawing a line under the means. Means not connected by underscoring with the same line are significantly different.

Coefficient of Variability:

$$\text{C.V.} = \sqrt{\frac{s^2}{\bar{x}_{..}}} \text{ x } 100 = \sqrt{\frac{29.442}{30.042}} \text{ x } 100$$

$$\text{C.V.} = \frac{5.42606}{30.042} \text{ x } 100 = \underline{18.1\%}$$

Expectations of the Mean Squares:

Table 4-4. Analysis of Variance Summary

Source	d.f.	EMS
f Treats	t-1	$\sigma^2 + r \sum_i^t \tau_i^2/t\text{-}1 \equiv \sigma^2 + [T]$
r Reps.	r-1	$\sigma^2 + t\sigma_\beta^2$
r Error	(t-1)(r-1)	σ^2

Comparison of Precision of RCB vs CRD:

$$\text{P.F.} = \frac{\text{MSE for CRD}}{\text{MSE for RCB}} \times 100\%$$

$$\text{MSE for CRD} = \frac{n_b E_b + (n_t + n_e) E_e}{n_t + n_b + n_e}$$

Where: n_b = d.f. for Reps. = 5
E_b = MS for Reps. = 16.442
n_t = d.f. for Treats. = 3
n_e = d.f. for Error = 15
E_e = MS for Error = 29.442

$$\text{MSE}_{CRD} = \frac{5(16.442) + (3 + 15)(29.442)}{3 + 5 + 15}$$

$$\text{MSE}_{CRD} = \frac{82.210 + 529.956}{23} = \underline{\underline{26.616}}$$

$$\text{P.F.} = \frac{26.616}{29.442} \times 100 = 90.4\%$$

The P.F. of 90.4% indicates that 100 observations using a RCB design will be needed to give the same precision (ability to detect small differences among treatment means) as 90.4 observations using the CRD.

PART TWO

Data:

Table 4-5. Analysis of Variance Summary

Source	d.f.	Mean Square	Expected Mean Square
r Reps.	4	240	$\sigma^2 + 2\sigma_2^2 + 12\sigma_3^2 + 72\sigma_\beta^2$
f Treats.	5	360	$\sigma^2 + 2\sigma_2^2 + 12\sigma_3^2 + 60 \sum_i^6 \tau_i^2/5$
r Exper. Error	20	120	$\sigma^2 + 2\sigma_2^2 + 12\sigma_3^2$
r Sampling Error	150	30	$\sigma^2 + 2\sigma_2^2$
r Deter: Samples	180	4	σ^2
Total	359		

Determination of Number of Treatments, Reps.:   c.:

Table 4-6. Summary of Number of Treatments

Source of Variation	d.f.	Number in Experiment
r Replication	$(r-1) = 4$	$r = 5$
f Treatments	$(t-1) = 5$	$t = 6$
r Experimental Error	$d.f._{Total}$ - all others	
r Sampling Error	$rt(s-1) = 150$	$s = 6$
r Determinations: Samples	$rts(d-1) = 180$	$d = 2$
Total	$(rtsd - 1) = 359$	$T = 360$

Decoding Data:

The mean square is essentially a variance, and coding the data would affect the mean square in the same manner it affects the variance.

Effects Due to Coding by Addition:

$$s_c^2 = \frac{\sum\limits_i^n [(x_i + k) - (\bar{x} + k)]^2}{n - 1} = \frac{\sum\limits_i^n (x_i - \bar{x})^2}{n - 1}$$

There is no effect on the variance by the addition of a constant; therefore, the value of the error mean square would remain the same when 10 was added to each determination.

Effects Due to Multiplication Coding:

$$s_c^2 = \frac{\sum\limits_i^n [x_i k - \bar{x} k]^2}{n - 1} = \frac{k^2 \sum (x_i - \bar{x})^2}{n - 1}$$

$$s_c^2 = k^2 s^2$$

$$s^2 = \frac{s_c^2}{k^2} = \frac{120}{(10)^2} = \frac{120}{100} = \underline{\underline{1.20}}$$

Computation of the Variance of a Treatment Mean:

$$s_{\bar{x}}^2 = \frac{\sigma^2 + 2\sigma_2^2 + 12\sigma_3^2}{rsd} = \frac{120}{60}$$

$$s_{\bar{x}}^2 = \underline{\underline{2.00}}$$

Tukey's w-procedure utilizes $s_{\bar{x}}$ which is easily obtained from the variance of a treatment mean by extracting the square root.

$$s_{\bar{x}} = \sqrt{s_{\bar{x}}^2} = \sqrt{2.00} = \underline{\underline{1.414}}$$

$$w = s_{\bar{x}} q_{a:p,n_2}$$

Estimate of the Gain or Loss in Efficiency if 8 Samples Per Plot are Selected with One Determination:

The change in precision by altering the sampling and determinations can be estimated by evaluating the variances of the treatment means for each situation and computing the precision factor.

$$S_{\bar{x}_t}^{2} = \frac{\sigma^2 + d\sigma_2^{\;2} + sd\sigma_3^{\;2}}{rsd}$$

For 6 samples and 2 determinations:

$$S_{\bar{x}_t}^{2} = \underline{\underline{2.00}}$$

$$\sigma^2 + 2\sigma_2^{\;2} = 30$$

$$4 + 2\sigma_2^{\;2} = 30$$

$$\sigma_2^{\;2} = \underline{\underline{13}}$$

$$\sigma^2 + 2\sigma_2^{\;2} + 12\sigma_3^{\;2} = 120$$

$$4 + 2(13) + 12\sigma_3^{\;2} = 120$$

$$12\sigma_3^{\;2} = 90$$

$$\sigma_3^{\;2} = \underline{\underline{7.50}}$$

Assuming the variances remain constant:

For 8 samples and 1 determination:

$$S_{\bar{x}_t}^{2} = \frac{4 + 13 + 8(7.50)}{40} = \frac{77}{40} = \underline{\underline{1.925}}$$

$$P.F. = \frac{S_{\bar{x}_t}^{2} \;(6\;\&\;2)}{S_{\bar{x}_t}^{2} \;(8\;\&\;1)} \; x\;100 = \frac{2.00}{1.925}\; x\;100 = \underline{\underline{103.9\%}}$$

A P.F. of 103.9% indicates that 104 observations would be required with the 6 samples and two determinations to give the same precision as 100 observations using the eight samples and one determination.

Conclusions:

All of the insecticide means differed significantly from the control; however, there was no significant difference between the insecticides used. Tukey's w-procedure gives the most conservative test for differences. With other testing procedures, it would have been shown that the Lindane 25% treatment was superior to the others; however, in this case it must be concluded that all insecticides were better than the control and one insecticide would be as effective as the others.

Exercise Number 5

The Latin Square Design

Problem:

A Latin square design was used at the University of Hawaii to compare 6 different legume intercycle crops for pineapples. The yields in 10-gram units are given in the accompanying table.

						Row Totals
B	F	D	A	E	C	
220	98	149	92	282	169	1010
A	E	B	C	F	D	
74	238	158	228	48	188	934
D	C	F	E	B	A	
118	279	118	278	176	65	1034
E	B	A	D	C	F	
295	222	54	104	213	163	1051
C	D	E	F	A	B	
187	90	242	96	66	122	803
F	A	C	B	D	E	
90	124	195	109	79	211	808
Col. Total	984	1051	916	907	864	918	5640

1. Make a complete analysis of this experiment.
2. Write out the expectations of the mean square assuming that the treatments are fixed and that the rows and columns are random variables.

3. Compare the efficiency of this Latin square design with that of RCB where the rows are used as replications and also where columns are used as replications. What is the relative efficiency of a completely random design?
4. Use the Student-Newman-Keuls' test to determine real differences among the treatment means.

SELECTED STANDARD LATIN SQUARES

3 x 3			4 x 4		
	1	2	3	4	

3 x 3

```
A B C
B C A
C A B
```

1

```
A B C D
B A D C
C D B A
D C A B
```

2

```
A B C D
B C D A
C D A B
D A B C
```

4 x 4

3

```
A B C D
B D A C
C A D B
D C B A
```

4

```
A B C D
B A D C
C D A B
D C B A
```

5 x 5

```
A B C D E
B A E C D
C D A E B
D E B A C
E C D B A
```

6 x 6

```
A B C D E F
B F D C A E
C D E F B A
D A F E C B
E C A B F D
F E B A D C
```

7 x 7

```
A B C D E F G
B C D E F G A
C D E F G A B
D E F G A B C
E F G A B C D
F G A B C D E
G A B C D E F
```

8 x 8

```
A B C D E F G H
B C D E F G H A
C D E F G H A B
D E F G H A B C
E F G H A B C D
F G H A B C D E
G H A B C D E F
H A B C D E F G
```

9 x 9

```
A B C D E F G H I
B C D E F G H I A
C D E F G H I A B
D E F G H I A B C
E F G H I A B C D
F G H I A B C D E
G H I A B C D E F
H I A B C D E F G
I A B C D E F G H
```

10 x 10

```
A B C D E F G H I J
B C D E F G H I J A
C D E F G H I J A B
D E F G H I J A B C
E F G H I J A B C D
F G H I J A B C D E
G H I J A B C D E F
H I J A B C D E F G
I J A B C D E F G H
J A B C D E F G H I
```

11 x 11

```
A B C D E F G H I J K
B C D E F G H I J K A
C D E F G H I J K A B
D E F G H I J K A B C
E F G H I J K A B C D
F G H I J K A B C D E
G H I J K A B C D E F
H I J K A B C D E F G
I J K A B C D E F G H
J K A B C D E F G H I
K A B C D E F G H I J
```

12 x 12

```
A B C D E F G H I J K L
B C D E F G H I J K L A
C D E F G H I J K L A B
D E F G H I J K L A B C
E F G H I J K L A B C D
F G H I J K L A B C D E
G H I J K L A B C D E F
H I J K L A B C D E F G
I J K L A B C D E F G H
J K L A B C D E F G H I
K L A B C D E F G H I J
L A B C D E F G H I J K
```

A standard Latin Square is one where the letters in the first column and first row are ordered.

Purpose:

To compare 6 different legume intercycle crops for pineapples which were tested at the University of Hawaii.

Methods and Materials:

A Latin square design was used to evaluate the data by the analysis of variance technique. The data were recorded as yields in 10-gram units of pineapples.

Results and Computations:

Data:

Table 5-1. Pineapple Yields, Treatment Totals and Means

Yields						Row Totals	Treat. Totals	Treat. Means
B 220	F 98	D 149	A 92	E 282	C 169	1010	A 475	A 79.17
A 74	E 238	B 158	C 228	F 48	D 188	934	B 1007	B 167.83
D 118	C 279	F 118	E 278	B 176	A 65	1034	C 1271	C 211.83
E 295	B 222	A 54	D 104	C 213	F 163	1051	D 728	D 121.33
C 187	D 90	E 242	F 96	A 66	B 122	803	E 1546	E 257.67
F 90	A 124	C 195	B 109	D 79	E 211	808	F 613	F 102.17
Col. Tot. 984	1051	916	907	864	918	5640		

Analysis of Variance Calculations:

Treatment Sum of Squares:

$$\text{TrSS} = \sum_{k}^{t} X_{t.}^{2}/t - X_{..}^{2}/t^{2}$$

$$TrSS = \frac{(475^2 + 1007^2 + 1271^2 + 728^2 + 1546^2 + 613^2)}{6} - \frac{5640^2}{36}$$

$$TrSS = \frac{6,150,984}{6} - \frac{31,809,600}{36}$$

$$TrSS = 1,025,164.0 - 883,600.0 = \underline{141,564}$$

Column Sum of Squares:

$$CSS = \sum_{j}^{t} X_{.j}^{2}/t - C.F.$$

$$CSS = \frac{(984^2 + 1051^2 + 916^2 + 907^2 + 864^2 + 918^2)}{6} - C.F.$$

$$CSS = \frac{5,323,782}{6} - 883,600$$

$$CSS = 887,297 - 883,600 = \underline{3,697}$$

Row Sum of Squares:

$$RSS = \sum_{i}^{t} X_{i.}^{2}/t - C.F.$$

$$RSS = \frac{(1010^2 + 934^2 + 1034^2 + 1051^2 + 803^2 + 808^2)}{6} - C.F.$$

$$RSS = \frac{5,363,886}{6} - 883,600$$

$$RSS = 893,981 - 883,600 = \underline{10,381}$$

Total Sum of Squares:

$$TSS = \sum_{i}^{t} \sum_{j}^{t} X_{ij}^{2} - C.F.$$

$$TSS = (220^2 + ... + 211^2) - 883,600$$

$$TSS = 1,067,856 - 883,600 = \underline{184,256}$$

Error Sum of Squares:

$$ESS = TSS - (TrSS + CSS + RSS)$$

$$ESS = 184,256 - (141,564 + 3,697 + 10,381)$$

ESS = 28,614

Mean Squares:

MSTr = TrSS/d.f. = 141,564/5

MSTr = 28,312.80

MSC = CSS/d.f. = 3,697/5

MSC = 739.40

MSR = RSS/d.f. = 10,381/5

MSR = 2,076.20

MSE = ESS/d.f. = 28,614/20

MSE = 1,430.70

Calculated F-value:

$$F_{Treats} = \frac{MST}{MSE} = \frac{28,312.80}{1,430.70} = 19.789**$$

Critical F-value From Table A-2:

$$F_{(5,20).05} = 2.71$$
$$_{.01} = 4.10$$

There exists a highly significant difference between the treatment means.

Table 5-2. Analysis of Variance Summary

Source	d.f.	SS	MS	F
Treats.	5	141,564	28,312.80	19.789**
Columns	5	3,697	739.40	
Rows	5	10,381	2,076.20	
Error	20	28,614	1,430.70	
Total	35	184,256		

Standard Error of the Means:

$$s_{\bar{x}} = \sqrt{\frac{s^2}{r}} \quad \text{where } s^2 = MSE$$

$$s_{\bar{x}} = \sqrt{\frac{1{,}430.70}{6}} = \sqrt{238.45} = \underline{\underline{15.44}}$$

Comparison of Efficiency of Latin Square with RCB Using Columns as Replications:

$$s_{RCB_c}^2 = \frac{n_C E_C + (n_t + n_e) E_e}{n_C + n_t + n_e} \quad , \text{ where}$$

$$
\begin{aligned}
n_C &= \text{d.f. for columns} \\
E_C &= MSC \\
n_t &= \text{d.f. for treatments} \\
n_e &= \text{d.f. for error for Latin Square} \\
E_e &= \text{MSE for Latin Square}
\end{aligned}
$$

$$s_{RCB_c}^2 = \frac{5(739.40) + (5+20)(1{,}430.70)}{5 + 5 + 20}$$

$$s_{RCB_c}^2 = \frac{3{,}697.00 + 35{,}796.50}{30} = \frac{39{,}464.50}{30}$$

$$s_{RCB_c} = \underline{\underline{1{,}315.4833}}$$

$$P.F. = \frac{s_{RCB_c}^2}{s_{LS}^2} \times 100 = \frac{1{,}315.4833}{1{,}430.70} \times 100$$

$$P.F. = \underline{\underline{91.95\%}}$$

The P.F. indicates that 92 observations in a RCB analysis with columns as replications would give an equivalent amount of information as a Latin square with 100 observations.

Comparison of Efficiency of Latin Square with RCB Using Rows as Replications:

$$s_{RCB_R}^2 = \frac{n_R E_R + (n_t + n_e) E_e}{n_R + n_t + n_e}$$

$$s_{RCB_R}^2 = \frac{5(2,076.20) + (5 + 20)(1430.70)}{5 + 5 + 20}$$

$$s_{RCB_R}^2 = \frac{10,381.0 + 35,767.50}{30} = \frac{46,148.5}{30}$$

$$s_{RCB_R}^2 = \underline{1,538.28}$$

$$P.F. = \frac{s_{RCB_R}^2}{s_{LS}^2} \times 100 = \frac{1,538.28}{1,430.70} \times 100$$

$$P.F. = \underline{\underline{107.52\%}}$$

The P.F. indicates that 108 observations would be required with a RCB analysis with rows as replications to yield an equal amount of information from a Latin square design of 100 observations.

Comparison of Efficiency of Latin Square with a CRD Where $n_t = n_r = n_c = n$:

$$s_{CRD}^2 = \frac{n-1 (E_r + E_c) + [(n-1) + (n-1)(n-2)] E_e}{2(n-1) + (n-1) + (n-1)(n-2)}$$

$$s_{CRD}^2 = \frac{5(2,076.20 + 739.40) + [5 + (5)(4)](1430.70)}{2(5) + 5 + 5(4)}$$

$$s_{CRD}^2 = \frac{14,078.0 + 35,767.5}{35} = \underline{\underline{1,424.16}}$$

$$P.F. = \frac{s_{CRD}^2}{s_{LS}^2} \times 100 = \frac{1424.16}{1430.70} \times 100 = \underline{\underline{99.54\%}}$$

The precision factor indicates that 100 observations would be required with a CRD analysis to yield an equal amount of information from a Latin square design of 100 observations. In other words, the Latin square and the CRD are equal in precision.

Comparison of Means by the Student-Newman-Keul's Test:

Table 5-3. Ranked Treatment Means

A	F	D	B	C	E
79.17	102.17	121.33	167.83	211.83	257.67

$W_p = q_a (p, n_2) \, s_{\bar{x}}$

q_a is obtained from Table A-4

p = number of means

n_2 = error degrees of freedom

$s_{\bar{x}}$ = standard error of a treatment mean

$s_{\bar{x}}$ = 15.44

Table 5-4. Student-Newman-Keul's Significant Ranges

p	2	3	4	5	6
$q_{.05(p,20)}$	2.95	3.58	3.96	4.23	4.45
$W_{p_{.05}}$	45.55	55.28	61.14	65.31	68.71

E vs A
257.67
-79.17
178.50 > 68.71

E vs F
257.67
-102.17
155.50 > 65.31

E vs D
257.67
-121.33
136.34 > 61.14

E vs B
257.67
-167.83
89.84 > 55.28

E vs C
257.67
-211.83
45.84 > 45.55

C vs A
211.83
-79.17
132.66 > 65.31

C vs F
211.83
-102.17
109.66 > 61.14

C vs D
211.83
-121.33
90.50 > 55.28

C vs B
211.83
-167.83
44.00 < 45.55

B vs A	B vs F	B vs D
167.83	167.83	167.83
-79.17	-102.17	-121.33
88.66 > 61.14	65.66 > 55.28	46.50 > 45.55

D vs A	F vs A
121.33	102.17
-79.17	-79.17
42.16 < 55.28	23.00 < 45.55

Means that do not differ significantly are joined by underscoring the means listed in Table 5-3.

Table 5-5. Expectations of the Mean Square Assuming That The Treatments are Fixed and Rows and Columns are Random Variables

	Source	d.f.	Expected Mean Square
f	Treats.	5	$\sigma^2 + r(\Sigma\tau_t^2)/r\text{-}1$
r	Columns	5	$\sigma^2 + r\sigma_c^2$
r	Rows	5	$\sigma^2 + r\sigma_R^2$
r	Error	20	σ^2
	Total	35	

Conclusions:

There are significant differences in the legume intercycles crops with method E superior to the others based upon the production of greater quantities of pineapple. Methods B and C are better than the remaining techniques. This experiment derived very little benefit from a Latin square design. The experiment would have been more efficient if it had been designed as a RCB experiment with columns as replications. There was a slight gain by using the Latin square design over the RCB design with rows as replications. Equal efficiency could have been gained using a CRD.

Exercise Number 6

Factorial Experiment 2 x 2

Problem:

A 2 x 2 factorial arranged in a CRD. Data supplied by Edith B. Davis, Microbiology Dept., Miss. State University.

Total nitrogen content, as determined by the semi-micro Kjeldahl method in mg, of algae grown in two different volumes of media, in atmospheres of hydrogen and nitrogen gases.

40 ml.		60 ml	
$CO_2 + N_2$	$CO_2 + H_2$	$CO_2 + N_2$	$CO_2 + H_2$
0.613	0.317	0.514	0.216
0.668	0.069	0.586	0.104
0.581	0.069	0.310	0.214
0.624	0.259	0.888	0.104
0.460	0.627	1.210	0.104
0.501	0.380	1.199	0.104

1. Write out the linear additive model and state the Null Hypotheses used.
2. Determine the composition of the expected mean squares when all of the factorial effects are fixed.
3. Calculate the sums of squares of A within each level of B and show that the sum of these equals the sums of squares for A + AB.
4. Calculate the coefficient of variability.
5. Use Keuls' method to detect real differences among marginal and/or subclass means.
6. Make a creditable interpretation of the results.

Purpose:

To determine if significant differences exist when algae are grown in different volumes of medium with two gas mixtures above the culture. The two gas mixtures were carbon dioxide and nitrogen and carbon dioxide and hydrogen.

Method and Materials:

A 2 x 2 factorial arranged in a completely random design was used to evaluate the data. The data were recorded as milligrams of total nitrogen as measured by the semi-micro Kjeldahl method.

Results and Computations:

Linear Additive Model:

$$X_{ijk} = \mu + a_i + \beta_j + (\alpha\beta)_{ij} + \epsilon_{ijk}$$

$$i = 1, \dots a = 2$$
$$j = 1, \dots b = 2$$
$$k = 1, \dots r = 6$$

$$H_o : a_i = \beta_j = (\alpha\beta)_{ij} = 0$$

Expected Mean Squares With All Factorial Effects Fixed:

Source	d.f.	EMS	F
f Vols	a-1	$z = \sigma^2 + rb \sum_i^a a_i^2 / a\text{-}1$	z/s
f Gases	b-1	$y = \sigma^2 + ra \sum_j^b \beta_j^2 / b\text{-}1$	y/s
f V x G	(a-1)(b-1)	$x = \sigma^2 + r \sum_i^a \sum_j^b (\alpha\beta)_j / (a\text{-}1)(b\text{-}1)$	x/s
r Reps.	r-1	$\sigma^2 + ab\sigma_e^2$	
r Error	(r-1)(ab-1)	$s = \sigma^2$	
Total	abr - 1		

Data:

Table 6-1. Total Nitrogen Content of Algae, mg

Volume	40 ml.	40 ml.	60 ml.	60 ml.
Gases	$CO_2 + N_2$	$CO_2 + H_2$	$CO_2 + N_2$	$CO_2 + H_2$
1	0.613	0.317	0.514	0.216
2	0.668	0.069	0.586	0.104
3	0.581	0.069	0.310	0.214
4	0.624	0.259	0.888	0.104
5	0.460	0.627	1.210	0.104
6	0.501	0.380	1.199	0.104

Table 6-2. Summary Table of Totals

	40 ml.	60 ml.	Total
N_2	3.447	4.707	8.154
H_2	1.721	0.846	2.567
Total	5.168	5.553	10.721

Computation of Sums of Squares:

The following formulas are applicable only to cases with single degrees of freedom.

Sum of Squares for Volumes:

$$VSS = \frac{(X_{1..} - X_{2..})^2}{abr} = \frac{(5.168 - 5.553)^2}{24}$$

$$VSS = \frac{(0.385)^2}{24} = \frac{0.148225}{24} = \underline{\underline{0.006176}}$$

Sum of Squares for Gases:

$$GSS = \frac{(X_{.1.} - X_{.2.})^2}{abr} = \frac{(8.154 - 2.567)^2}{24}$$

$$GSS = \frac{(5.587)^2}{24} = \frac{31.214569}{24} = \underline{\underline{1.300607}}$$

Sum of Squares for Totals:

$$TSS = \sum_i^a \sum_j^b \sum_k^r X_{ijk}^2 - X_{...}^2 / abr$$

$$TSS = 7.255805 - (10.721)^2 / 24$$

$$TSS = 7.255805 - 114.939841 / 24$$

$$TSS = 7.255805 - 4.789160$$

$$TSS = \underline{\underline{2.466645}}$$

Sum of Squares for Interaction:

$$ISS = \frac{[(X_{11.} + X_{22.}) - (X_{21.} + X_{12.})]^2}{4r}$$

$$ISS = \frac{[(3.447 + 0.846) - (4.707 + 1.721)]^2}{4(6)}$$

$$ISS = \frac{(4.293 - 6.428)^2}{24} = \frac{(2.135)^2}{24}$$

$$ISS = \frac{4.558225}{24} = \underline{\underline{0.189926}}$$

Table 6-3. Analysis of Variance Summary

Source	d.f.	S.S.	M.S.	F
Vols.	1	0.006176	0.006176	< 1 n.s.
Gases	1	1.300607	1.300607	26.82**
V x G	1	0.189926	0.189926	3.92 n.s.
Error	20	0.969936	0.048497	
Total	23	2.466645		

Tabular Value of F from Table A-2:

$$F_{(1,20).05} = 4.35$$
$$F_{(1,20).01} = 8.10$$

Comparison with calculated F-value is shown in Analysis of Variance Summary Table by the usual symbols, i.e. n.s. = non-significant, * = significant and ** = highly significant.

Calculation of Sums of Squares of A Within Each Level of B:

$$SS_{AwB} = \frac{(X_{21.} - X_{11.})^2}{ra} + \frac{(X_{12.} - X_{22.})^2}{ra}$$

$$SS_{AwB} = \frac{(4.707 - 3.447)^2}{12} + \frac{(1.721 - 0.846)^2}{12}$$

$$SS_{AwB} = 0.132300 + 0.063802$$

$$SS_{AwB} = \underline{\underline{0.196102}}$$

$$SS_{AwB} = \text{should equal A + AB}$$

$$SSA = 0.006176$$

$$SSAB = 0.189926$$

$$SSA + SSAB = \underline{\underline{0.196102}}$$

Coefficient of Variability:

$$C.V. = \frac{\sqrt{s^2}}{\bar{x}} \times 100 = \frac{\sqrt{0.048497}}{10.721/24} \times 100$$

$$C.V. = \frac{0.2202203}{0.446708} \times 100 = \underline{\underline{49.3\%}}$$

Keul's Method To Detect Real Differences:

When an interaction is not significant, marginal means can be compared for significance because the factors act independently in producing the response.

Table 6-4. Summary Table of Means

	40 ml.	60 ml.	Mar. Means
N_2	0.5745	0.7845	0.6795
H_2	0.2868	0.1410	0.2139
Mar. Means	0.4307	0.4628	0.4467

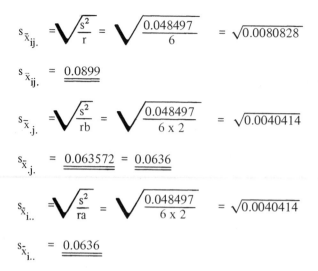

$$s_{\bar{x}_{ij.}} = \sqrt{\frac{s^2}{r}} = \sqrt{\frac{0.048497}{6}} = \sqrt{0.0080828}$$

$$s_{\bar{x}_{ij.}} = \underline{\underline{0.0899}}$$

$$s_{\bar{x}_{.j.}} = \sqrt{\frac{s^2}{rb}} = \sqrt{\frac{0.048497}{6 \times 2}} = \sqrt{0.0040414}$$

$$s_{\bar{x}_{.j.}} = 0.063572 = \underline{\underline{0.0636}}$$

$$s_{\bar{x}_{i..}} = \sqrt{\frac{s^2}{ra}} = \sqrt{\frac{0.048497}{6 \times 2}} = \sqrt{0.0040414}$$

$$s_{\bar{x}_{i..}} = \underline{\underline{0.0636}}$$

$$W_p = q_a (p, n_2)\, s_{\bar{x}}$$

q_a is obtained from Table A-4

p = number of means = 2

n_2 = error degrees of freedom = 20

$s_{\bar{x}}$ = standard error of a mean

$q_{.05(2,20)}$ = 2.95

$q_{.01(2,20)}$ = 4.02

$W_{p_{.05\ \&\ .01}}$ for $s_{\bar{x}_{.j.}}$:

$$W_{p_{.05}} = 2.95(0.0636) = 0.1876$$

$$W_{p_{.01}} = 4.02(0.0636) = 0.2557$$

$W_{p_{.05\ \&\ .01}}$ for $s_{\bar{x}_{i..}}$:

$$W_{p_{.05}} = 2.95(0.0636) = 0.1876$$

$$W_{p_{.01}} = 4.02(0.0636) = 0.2557$$

Comparison of $\bar{x}_{1..}$ **vs** $\bar{x}_{2..}$:

$\bar{x}_{1..} = 0.4307, \bar{x}_{2..} = 0.4628$

$\bar{x}_{2..} - \bar{x}_{1..} = 0.4628 - 0.4307 = \underline{\underline{0.0321}}$

$0.0321 < 0.1876$ ∴ no significant difference.

Comparison of $\bar{x}_{.1.}$ **vs** $\bar{x}_{.2.}$:

$\bar{x}_{.1.} = 0.6795, \bar{x}_{.2.} = 0.2139$

$\bar{x}_{.1.} - \bar{x}_{.2.} = 0.6795 - 0.2139 = 0.4656$

$0.4656 > 0.2557$ ∴ a highly significant difference.

Conclusions:

The comparison of marginal means indicates that the difference in volume of medium did not affect the experiment. However, a large difference exists due to the variation in the gas content of the atmosphere above the culture.

Due to the very high coefficient of variability it is doubtful that the above interpretations are justified. Variability of this type is often encountered in biological studies and is unexplainable.

Exercise Number 7

Split-Plot Design

Problem:

(Data supplied by E. L. Moore and R. R. Bruce, Miss. State Univ., 1957.)

Temperature readings in degrees Fahrenheit (averaged over a period of four different days) at different positions and/or soil depths for various mulch treatments applied to a tomato crop.

Type of Tomato Mulch	Thermometer Positions	Replications		
		1	2	3
Soil Mulch (tomatoes trailing)	On surface	129	134	128
	Underneath	100	104	107
	2" deep in soil	94	94	98
	4" deep in soil	89	92	93
	6" deep in soil	88	88	90
		500	512	516
Soil Mulch (tomatoes staked)	On surface	130	134	130
	Underneath	98	103	104
	2" deep in soil	93	95	97
	4" deep in soil	89	91	92
	6" deep in soil	88	89	90
		498	512	513

Type of Tomato Mulch	Thermometer Positions	Replications		
		1	2	3
Black Plastic (tomatoes staked)	On surface	136	139	136
	Underneath	112	116	114
	2" deep in soil	100	109	108
	4" deep in soil	96	102	100
	6" deep in soil	91	95	96
		535	561	554
White Plastic (tomatoes staked)	On surface	110	112	111
	Underneath	102	108	107
	2" deep in soil	99	104	104
	4" deep in soil	93	97	98
	6" deep in soil	89	92	92
		493	513	512
Straw (tomatoes staked)	On surface	133	139	138
	Underneath	88	90	88
	2" deep in soil	81	81	82
	4" deep in soil	80	80	81
	6" deep in soil	79	80	80
		461	470	469
Sawdust (tomatoes staked)	On surface	134	142	138
	Underneath	86	87	87
	2" deep in soil	82	83	83
	4" deep in soil	81	82	82
	6" deep in soil	80	81	82
		463	475	472
Total		2950	3043	3036

Purpose:

To evaluate the effects of various mulches upon the temperature of the soil at various depths around tomato plants.

Methods and Materials:

A split-plot design was used to evaluate the temperature data. The data were recorded in degrees Fahrenheit, averaged over a period of four days, at different positions and soil depths for various mulch treatments applied to a tomato crop.

Results and Computations:

Data: The raw data are shown above.

Summary Table of Totals:

Factor B-Depths

Factor A-Mulches	b_1	b_2	b_3	b_4	b_5	Marginal Total
Soil (Standing)	391	311	286	274	266	1528
Soil (Staked)	394	305	285	272	267	1523
Black Plastic (Staked)	411	342	317	298	282	1650
White Plastic (Staked)	333	317	307	288	273	1518
Straw (Staked)	410	266	244	241	239	1400
Sawdust (Staked)	414	260	248	245	243	1410
Marginal Totals	2353	1801	1687	1618	1570	9029

b_1 = On surface
b_2 = Underneath
b_3 = 2'' deep
b_4 = 4'' deep
b_5 = 6'' deep

Analysis of Variance Calculations:

Total Sum of Squares:

$$\text{TSS} = \sum_{i}^{a} \sum_{j}^{r} \sum_{k}^{b} X_{ijk}^2 - X_{...}^2/arb$$

$$\text{TSS} = (129^2 + 134^2 + \ldots + 82^2) - 9029^2/90$$

$$\text{TSS} = \underline{\underline{28{,}839.7}}$$

Mulch Sum of Squares:

$$\text{MSS} = \sum_{i}^{a} X_{i..}^2/rb - \text{C.F.}$$

$$\text{MSS} = \frac{(1528^2 + 1523^2 + 1650^2 + 1518^2 + 1400^2 + 1410^2)}{3(5)} - \text{C.F.}$$

$$\text{MSS} = 13,629,237/15 - 905,809.3$$

$$\text{MSS} = 908,615.8 - 905,809.3 = \underline{\underline{2,806.5}}$$

Replications Sum of Squares:

$$\text{RSS} = \sum_{j}^{r} X_{.j.}^{2}/ab - \text{C.F.}$$

$$\text{RSS} = \frac{(2950^2 + 3043^2 + 3036^2)}{6(5)} - 905,809.3$$

$$\text{RSS} = 27,179,645/30 - 905,809.3$$

$$\text{RSS} = 905,988.1 - 905,809.3 = \underline{\underline{178.8}}$$

Error a Sum of Squares:

$$E_a \text{SS} = \sum_{i}^{a} \sum_{j}^{r} C_{ij.}^{2}/b - \text{RSS} - \text{MSS} - \text{C.F.}$$

$$E_a \text{SS} = \frac{(500^2 + ... + 472^2)}{5} - 2,806.5 - 178.8 - 905,809.3$$

$$E_a \text{SS} = 4,544,101/5 - 2,806.5 - 178.8 - 905,809.3$$

$$E_a \text{SS} = 908,820.2 - 2,806.5 - 178.8 - 905,809.3$$

$$E_a \text{SS} = \underline{\underline{25.60}}$$

Positions Sum of Squares:

$$\text{PSS} = \sum_{k}^{b} X_{..k}^{2}/ra - \text{C.F.}$$

$$\text{PSS} = \frac{(2353^2 + 1801^2 + 1687^2 + 1618^2 + 1570^2)}{3(6)} - \text{C.F.}$$

$$\text{PSS} = 16,709,003/18 - 905,809.3$$

$$\text{PSS} = 928,277.9 - 905,809.3 = \underline{\underline{22,468.6}}$$

Positions x Mulches Sum of Squares:

$$\text{PxMSS} = \sum_{i}^{a} \sum_{k}^{b} X_{i.k}^{2}/r - \text{C.F.} - \text{MSS} - \text{PSS}$$

$$\text{PxMSS} = \frac{(391^2 + 311^2 + ... + 243^2)}{3} - 905,809.3 - 2,806.5 - 22,468.6$$

$$\text{PxMSS} = 2,802,959/3 - 905,809.3 - 2,806.5 - 22,468.6$$

$$\text{PxMSS} = 934,319.6 - 905,809.3 - 2,806.5 - 22,468.6$$

$$\text{PxMSS} = \underline{3,235.2}$$

Error b Sum of Squares:

$$E_b \text{SS} = \text{TSS} - \text{All other SS}$$

$$E_b \text{SS} = 28,839.7 - 2,806.5 - 178.8 - 25.6 - 22,468.6 - 3,235.2$$

$$E_b \text{SS} = \underline{125.0}$$

Mean Squares:

$$\text{MSM} = \frac{\text{MSS}}{\text{d.f.}} = 2.806.5/5 = \underline{\underline{561.3}}$$

$$\text{MSR} = \frac{\text{RSS}}{\text{d.f.}} = 178.8/2 = \underline{\underline{89.4}}$$

$$\text{MSE}_a = \frac{E_a \text{SS}}{\text{d.f.}} = 25.6/10 = 2.56 = \underline{\underline{2.6}}$$

$$\text{MSP} = \frac{\text{PSS}}{\text{d.f.}} = 22,468.6/4 = \underline{\underline{5,617.2}}$$

$$\text{MSPxM} - \frac{\text{PxMSS}}{\text{d.f.}} = 3,235.2/20 = \underline{\underline{161.8}}$$

$$\text{MSE}_b = \frac{E_b \text{SS}}{\text{d.f.}} = 125.0/48 = 2.604 = \underline{\underline{2.6}}$$

Tabular Critical Values of F from Table A-2:

$$F_{.05}^{} {(5,10)} = 3.33$$
$$_{.01} = 5.64$$

$$F_{.05}^{} {(4,48)} = 2.58$$
$$_{.01} = 3.76$$

$$F_{.05}^{} {(20,48)} = 1.80$$
$$_{.01} = 2.30$$

Calculated F-values:

$$F_M = \frac{561.3}{2.6} = \underline{\underline{215.9**}}$$

$$F_P = \frac{5,617.2}{2.6} = \underline{\underline{2,160.5**}}$$

$$F_{PxM} = \frac{161.8}{2.6} = \underline{\underline{62.2**}}$$

Analysis of Variance Summary Table:

	Source	d.f.	S.S.	MS	F
f	Mulches	5	2,806.5	561.3	215.9**
r	Reps.	2	178.8	89.4	
r	E_a	10	25.6	2.6	
f	Positions	4	22,468.6	5,617.2	2,160.5**
f	PxM	20	3,235.2	161.8	62.2**
r	E_b	48	125.0	2.6	
	Total	89	28,839.7		

Comparison of Means:

Interaction (PxM) is highly significant; therefore, marginal means cannot be compared. It is necessary to compare sub-class means.

When comparing means and a weighted $s_{\bar{x}}$ on $s_{\bar{d}}$ is used, the calculation of (a) the LSD and (b) the multiple range tests are altered as follows:

(a) LSD

$$\text{LSD}_{.05} = s_{\bar{d}} \times t_{.05}'$$

$t_{.05}'$ is the critical value of t and will lie somewhere between the critical values of $t_{.05}$ corresponding to E_a and $t_{.05}$ corresponding to E_b.

$$t_{.05}' = \frac{(b\text{-}1)\, E_b t_{b.05} + E_a t_{a.05}}{(b\text{-}1)\, E_b + E_a}$$

b = number of levels of factor B

E_b = mean square for error b

$t_{b.05}$ = tabular value at a = .05 for df in E_b

E_a = mean square for error a

$t_{a.05}$ = tabular value at a = .05 for df in E_a

(b) Multiple Comparison Tests:

1. Calculate $s_{\bar{x}}$
2. Compute $t_{.05}'$
3. Insert the above numerical value of $t_{.05}'$ in the column of a table of t which corresponds to a = .05 probability and read off the corresponding degrees of freedom.
4. Use such df to read off the critical R_p or q_a values which are used for the various tests.

Duncan's New Multiple Range Test:

Comparisons Within Each Row of Summary Table:

Standard Error:

$$s_{\bar{x}} = \sqrt{\frac{E_b}{r}} = \sqrt{\frac{2.6}{3}} = \sqrt{0.867} = \underline{\underline{0.931}}$$

$$\text{LSR}_{.05} = s_{\bar{x}} \times \text{SR}_{.05\,(48)}$$

Value of p	2	3	4	5
$\text{SSR}_{.05}$	2.85	2.99	3.09	3.16
LSR	2.65	2.78	2.88	2.94

Summary Table of Means:

Factor B-Depths

		b_1	b_2	b_3	b_4	b_5	Marginal Means
Factor A-Mulches	a_1	130.3	103.7	95.3	91.3	88.7	101.9
	a_2	131.3	101.7	95.0	90.7	89.0	101.5
	a_3	137.0	114.0	105.7	99.3	94.0	110.0
	a_4	111.0	105.7	102.3	96.0	91.0	101.2
	a_5	136.7	88.7	81.3	80.3	79.7	93.3
	a_6	138.0	86.7	82.7	81.7	81.0	94.0
Marginal Means		130.7	100.1	93.7	89.9	87.2	100.3

a_1	= Soil (Standing)		b_1	= On surface
a_2	= Soil (Staked)		b_2	= Underneath
a_3	= Black Plastic (Staked)		b_3	= 2" deep
a_4	= White Plastic (Staked)		b_4	= 4" deep
a_5	= Straw (Staked)		b_5	= 6" deep
a_6	= Sawdust (Staked)			

Row Means Ranked:

	a_1b_5	a_1b_4	a_1b_3	a_1b_2	a_1b_1
Soil (Standing)	88.7	91.3	95.3	103.7	130.3
	a_2b_5	a_2b_4	a_2b_3	a_2b_2	a_2b_1
Soil (Staked)	89.0	90.7	95.0	101.7	131.3
	a_3b_5	a_3b_4	a_3b_3	a_3b_2	a_3b_1
Black Plastic (Staked)	94.0	99.3	105.7	114.0	137.0
	a_4b_5	a_4b_4	a_4b_3	a_4b_2	a_4b_1
White Plastic (Staked)	91.0	96.0	102.3	105.7	111.0
	a_5b_5	a_5b_4	a_5b_3	a_5b_2	a_5b_1
Straw (Staked)	79.7	80.3	81.3	88.7	136.7
	a_6b_5	a_6b_4	a_6b_3	a_6b_2	a_6b_1
Sawdust (Staked)	81.0	81.7	82.7	86.7	138.0

Means not connected by underscoring with the same line are significantly different.

Null Hypothesis:

$$\mu_1 = \mu_2 = \mu_3 = \mu_4 = \mu_5$$

Comparisons Within Each Column of Summary Table:

$$s_{\bar{x}} = \sqrt{\frac{1}{r}\left\{\frac{(b-1)E_b + E_a}{b}\right\}} = \sqrt{\frac{1}{3}\left\{\frac{4(2.6) + 2.6}{5}\right\}}$$

$$s_{\bar{x}} = \sqrt{\frac{1}{3}\left\{\frac{10.4 + 2.6}{5}\right\}} = \sqrt{0.867} = \underline{0.931}$$

$$t_{.05} = \frac{(b-1)E_b t_{b.05} + E_a t_{a.05}}{(b-1)E_b + E_a}$$

$$t_{b.05} = 2.010, \ t_{a.05} = 2.228$$

$$t_{.05}' = \frac{4(2.6)\,2.010 + 2.6(2.228)}{4(2.6) + 2.6} = \frac{20.904 + 5.793}{13.0}$$

$$t_{.05}' = \frac{26.69}{13} = \underline{2.054}$$

$$d.f. = \underline{\underline{26}}$$

Value of p	2	3	4	5	6
$SSR_{.05}$	2.91	3.06	3.14	3.21	3.27
$LSR_{.05}$	2.71	2.85	2.92	2.99	3.04

Column Means Ranked:

	a_4b_1	a_1b_1	a_2b_1	a_5b_1	a_3b_1	a_6b_1
Surface	111.0	130.3	131.3	136.7	137.0	138.0

	a_6b_2	a_5b_2	a_2b_2	a_1b_2	a_4b_2	a_3b_2
Underneath	86.7	88.7	101.7	103.7	105.7	114.0

	a_5b_3	a_6b_3	a_2b_3	a_1b_3	a_4b_3	a_3b_3
2" deep	81.3	82.7	95.0	95.3	102.3	105.7

	a_5b_4	a_6b_4	a_2b_4	a_1b_4	a_4b_4	a_3b_4
4" deep	80.3	81.7	90.7	91.3	96.0	99.3

	a_5b_5	a_6b_5	a_1b_5	a_2b_5	a_4b_5	a_3b_5
6" deep	79.7	81.0	88.7	89.0	91.0	94.0

Means not connected by underscoring with the same line are significantly different.

Conclusions:

The black plastic provides the best mulch as far as providing greater quantities of heat at depths below the surface of the soil. At the surface, straw and sawdust provided equal temperature increases with the black plastic, but the black plastic was superior as the depth increased. The straw and sawdust provided equal temperature increases below the surface. The two soil mulches gave equal increases in temperature. The white plastic ranked next to the black plastic below the surface but provided the poorest increase when compared at the surface.

Exercise Number 8

Linear Regression and Correlation

Problem:

1. Determine the regression coefficient by regression of Y upon X.
2. Test $b_{y.x}$ for significance by means of F and t tests. Show that the two tests are identical.
3. Calculate the standard error of estimate.
4. Determine the regression equation and use it to prepare the regression line.
5. Determine the 95% confidence limits for the regression coefficients and regression line. Enter the latter on the regression chart.
6. Determine the correlation coefficient.
7. Test it for significance by the F and t tests and show that these give identical conclusions as the corresponding tests for the regression coefficient.
8. Interpret data.

Regression (ungrouped data). Collins, E. V. Factors Influencing the Draft of Plows. Am. Soc. Agr. Eng., Trans. 14-15:39-55, 1920.

Force in pounds (Y) required to pull a tractor at varying speeds (X).

X	0.9	1.3	2.0	2.7	3.4	3.4	4.1	5.2	5.5	6.0
Y	425	420	480	495	540	530	590	610	690	680

Purpose:

To determine if there exists a significant linear regression between the force required to pull a tractor at varying speeds.

Method and Materials:

A linear regression analysis was used to analyze the data. The data for the force required to pull a tractor are recorded in pounds and the speeds are recorded in miles per hour.

Results and Computations:

Data:

X	Y	XY
0.9	425	382.5
1.3	420	546.0
2.0	480	960.0
2.7	495	1,336.5
3.4	540	1,836.0
3.4	530	1,802.0
4.1	590	2,419.0
5.2	610	3,172.0
5.5	690	3,795.0
6.0	680	4,080.0

$$\Sigma X = 34.5 \qquad \Sigma Y = 5460 \qquad \Sigma XY = 20,329.0$$

$$\bar{x} = \underline{3.45} \;,\; \bar{y} = \underline{546}$$

$$\Sigma y^2 = \Sigma Y^2 - \frac{(\Sigma Y)^2}{n} = 3,063,650 - 2,981,160 = \underline{82,490}$$

$$\Sigma x^2 = \Sigma X^2 - \frac{(\Sigma X)^2}{n} = 147.01 - 119.02 = \underline{27.99}$$

$$\Sigma xy = \Sigma XY - \frac{\Sigma X \Sigma Y}{n} = 20,329.0 - 18,837.0 = \underline{1,492.0}$$

Prior to any calculations the data were plotted on rectangular graph paper to be sure the relationship approximated a straight line. This is shown on Figure 8-1 along with the final results which were calculated as follows.

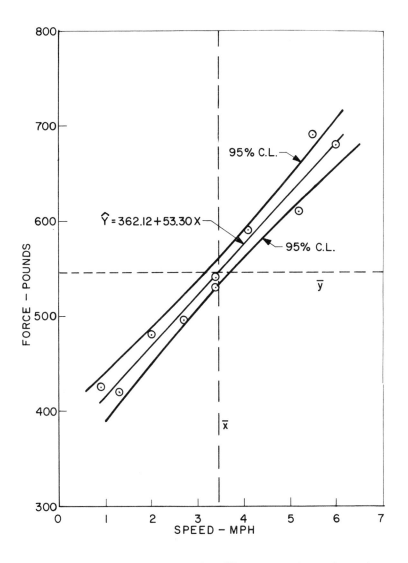

Figure 8-1. Plot of data showing results of linear regression and correlation.

Testing Significance of Regression:

Analysis of Variance Table

Source	d.f.	S.S.	MS	F
Regression	1	$(1492.0)^2/27.99 = 79{,}530.68$	79,530.68	214.99**
Residual	8	2,959.32	369.92	
Total	9	82,490		

The F-value is highly significant which indicates that Y is dependent upon the value of X, or that a linear relationship exists.

Prediction Equation:

$$\hat{Y} = \bar{y} + b_1 (X - \bar{x})$$

$$\hat{Y} = b_0 + b_1 X \text{ where } b_0 = \bar{y} - b_1 \bar{x}$$

$$b_1 = \frac{\Sigma xy}{\Sigma x^2} = \frac{1492.0}{27.99} = \underline{53.30}$$

$$b_0 = 546 - 53.30 (3.45) = 546 - 183.88 = \underline{362.12}$$

$$\hat{Y} = 362.12 + 53.30X$$

Computation of Points Used to Sketch the Prediction Equation:

When   X = 1.0
$$\hat{Y} = 362.12 + 53.30(1.0) = \underline{415.42}$$

When   X = 5.0
$$\hat{Y} = 362.12 + 53.30(5.0) = 362.12 + 266.50 = \underline{628.62}$$

These two predicted values are plotted on Figure 8-1 and joined by a straight line which will pass through the intersection of the $\bar{x}$ line and the $\bar{y}$ line and the y-intercept if arithmetic errors were not made.

Determination of 95% Confidence Limits for Regression Line:

$$\hat{Y} = \bar{y} + b_1 X$$

The variance of $\hat{Y}$ is equal to:

$$s_{\hat{Y}}^2 = \frac{s_{y.x}^2}{n} + \frac{s_{y.x}^2}{\Sigma x^2} x^2$$

$$s_{\hat{Y}} = \sqrt{s_{y.x}^2 \left\{ \frac{1}{n} + \frac{x^2}{\Sigma x^2} \right\}} \qquad \text{where } x = X - \bar{x}$$

$$s_{\hat{Y}} = \sqrt{369.92 \left\{ \frac{1}{10} + \frac{x^2}{27.99} \right\}}$$

95% C.L. for $\hat{Y}$ are: $\hat{Y} \pm t_{.05\,[n-2]} \times s_{\hat{Y}}$

95% C.L. for $\hat{Y}$ at $\bar{x}$ are:

$\hat{Y} = \bar{y}$

$\bar{y} \pm t_{.05} \times s_{\bar{y}} = 546.00 \pm 2.306 \sqrt{36.992}$

$546.00 \pm 2.306(6.082) = 546.00 \pm 14.02$

<u>560.02 to 531.98</u>

95% C.L. for $\hat{Y}$ at Values Other Than $\bar{x}$ Are:

Let $X = 1.0$

$$s_{\hat{Y}} = \sqrt{369.92 \left\{ \frac{1}{10} + \frac{(1 - 3.45)^2}{27.99} \right\}} = \sqrt{369.92(0.10 + 0.2145)}$$

$s_{\hat{Y}} = \sqrt{116.34} = \underline{10.7861}$

$\hat{Y} \pm t_{.05} s_{\hat{Y}} = 415.42 \pm 2.306 (10.7861)$

415.42 ± 24.87

<u>440.29 to 390.55</u>

Let $X = 5.90$, when $X = 5.90$, $(X - \bar{x})^2$ is the same as when $X = 1.0$; therefore,

$s_{\hat{Y}} = \underline{10.7861}$

$\hat{Y} = 362.12 + 53.30X = 362.12 + 53.30(5.90)$

$\hat{Y} = 362.12 + 314.47 = \underline{676.59}$

$\hat{Y} \pm t_{.05} s_{\hat{Y}} = 676.59 \pm 24.87$

$$\underline{\underline{701.46 \; \text{to} \; 651.72}}$$

Let X = 2.5

$$\hat{Y} = 362.12 + 53.30(2.5) = 362.12 + 133.25 = \underline{495.37}$$

$$s_{\hat{Y}} = \sqrt{369.92 \left\{ \frac{1}{10} + \frac{(2.5 - 3.45)^2}{27.99} \right\}} = \sqrt{369.92(0.13224)}$$

$$s_{\hat{Y}} = \sqrt{48.9182} = \underline{6.9942}$$

$$\hat{Y} \pm t_{.05} \, s_{\hat{Y}} = 495.37 \pm 2.306(6.9942)$$

$$495.37 \pm 16.13$$

$$\underline{\underline{511.50 \; \text{to} \; 479.24}}$$

Let X=4.40, when X=4.40, $(X - \bar{x})^2$ is the same as when X = 2.50; therefore:

$$s_{\hat{Y}} = 6.9942$$

$$\hat{Y} = 362.12 + 53.30(4.40) = 362.12 + 234.52$$

$$\hat{Y} = \underline{596.64}$$

$$\hat{Y} \pm t_{.05} \, s_{\hat{Y}} = 596.64 \pm 16.13$$

$$\underline{\underline{612.77 \; \text{to} \; 580.51}}$$

The above 95% C.L. were then plotted on Figure 8-1 and joined by a continuous line.

Testing Regression Coefficient by the t-test:

$$t = \frac{b_1 - \beta_1}{s_{b_1}}$$

$$H_0 : \beta_1 = 0$$

$$t = \frac{b_1}{s_{b_1}}$$

$$t_{\substack{.05 \\ .01}} [n\text{-}2] \; \begin{array}{l} = 2.306 \\ = 3.355 \end{array}$$ (From Table A-1)

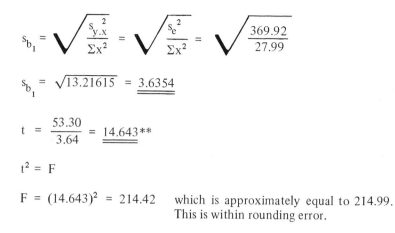

$$s_{b_1} = \sqrt{\frac{s_{y.x}^2}{\Sigma x^2}} = \sqrt{\frac{s_e^2}{\Sigma x^2}} = \sqrt{\frac{369.92}{27.99}}$$

$$s_{b_1} = \sqrt{13.21615} = \underline{3.6354}$$

$$t = \frac{53.30}{3.64} = \underline{14.643**}$$

$$t^2 = F$$

$F = (14.643)^2 = 214.42$ which is approximately equal to 214.99. This is within rounding error.

Determination of Correlation Coefficient:

$$r = \frac{\Sigma xy}{\sqrt{\Sigma y^2 \Sigma x^2}} = \frac{1492.0}{\sqrt{82,490 \times 27.99}} = \frac{1492.0}{\sqrt{2,308,895.1}}$$

$$r = \frac{1492.0}{1519.5} = \underline{0.9819}$$

Testing Significance of Correlation Coefficient:

To show that conclusions will be identical with tests for the regression coefficient:

Analysis of Variance Table

Source	d.f.	SS	MS	F
Reg.	1	$r^2 \Sigma y^2$	$r^2 \Sigma y^2 / 1$	MSR/MSE
Error	n-2	$(1 - r^2) \Sigma y^2$	$(1 - r^2) \Sigma y^2 / n-2$	
Total	n-1	Σy^2		

$$F = \frac{\dfrac{r^2 \Sigma y^2}{1}}{\dfrac{(1 - r^2) \Sigma y^2}{n - 2}} = \frac{r^2 \Sigma y^2 (n - 2)}{(1-r^2) \Sigma y^2} = \frac{r^2 (n - 2)}{(1 - r^2)}$$

$$t = \frac{r\sqrt{n-2}}{\sqrt{1-r^2}}$$

For 1 d.f. : $t = \sqrt{F}$

Analysis of Variance Table

Source	d.f.	SS	MS	F
Reg.	1	79,531.08	79,531.08	215.02**
Error	8	2,958.92	369.87	
Total	9	82,490.00		

F for the regression coefficient and the correlation coefficient agree within rounding errors:

$$214.99 \approx 215.02$$

$$t = \frac{0.9819\sqrt{10-2}}{\sqrt{1-0.9819^2}} = \frac{0.9819\,(2.8284)}{\sqrt{0.03587}} = \frac{2.77721}{0.1894}$$

$t = 14.663**$, which also agrees within rounding errors:

$$14.663 \approx 14.643$$

Conclusions:

There is a definite linear relationship between the force required to pull a tractor and the speed at which it is pulled. All tests indicate a highly significant relationship and examining the data plotted on linear graph paper also indicates such should be the case.

Exercise Number 9

Estimating Probability Model, Percentiles of Distribution, and Distribution Parameters with Probability Plotting

Problem:

1. Determine by using probability plotting if the data fit a normal distribution.
2. Determine the probability that the effluent total COD will not exceed a value of 60 mg/l:
3. Determine the mean, median and standard deviation.

A sewage treatment plant produced the following effluent quality as measured by the chemical oxygen demand (COD) test. Analyses were performed on the effluent and the filtrate of the effluent. Over a four-month sampling period the following values were obtained:

Sample	Collection Date	Chemical Oxygen Demand, mg/l	
		Effluent	Effluent Filtrate
1	2/03/67	56	25
2	3/08/67	63	36
3	3/11/67	57	36
4	3/18/67	33	23

Sample	Collection Date	Chemical Oxygen Demand, mg/l	
		Effluent	Effluent Filtrate
5	3/22/67	21	18
6	3/25/67	17	8
7	3/27/67	25	13
8	3/28/67	49	15
9	4/05/67	21	24
10	4/08/67	35	29
11	4/15/67	65	26
12	4/19/67	35	30
13	4/21/67	31	26
14	5/01/67	21	21
15	5/02/67	52	28
16	5/11/67	28	28
17	5/12/67	41	29
18	5/18/67	36	22

Purpose:

To determine by probability plotting if the sewage treatment plant effluent quality, as measured by the COD test, fits a normal distribution model, to estimate the probability of the effluent COD exceeding certain limits, and to estimate the distribution parameters, i.e. mean, median and standard deviation.

Method and Materials:

Normal probability paper was used to plot the ranked effluent quality values to determine if the plots fit a straight line. After establishing that the data approximate a straight line, it is possible to estimate the standard statistical parameters. The data used represent the effluent quality of a sewage treatment plant.

In the preceding exercises statistical tests have been used to provide more objective evaluations as to how well data fits a normal distribution. The results of these earlier tests are framed in a probabilistic manner and aid greatly in evaluating the adequacy of the model (normal distribution). Another very useful took in assessing the fit of data to a particular model is probability plotting. Probability plotting is subjective in that the determination of whether or not the data fit the assumed model is based on a visual examination. Probability plotting has many advantages in that it is expedient and simple as well as providing much useful information. It is frequently desirable to apply both the statistical testing as well as probability plotting. This is particularly true when the plotted data do not give a clear linear relationship. Then statistical calculations aid in determining the probability of the fit.

When employing probability plotting, it is best to select a model or type of probability paper that is based on an understanding of the physical phenomena. If there is a total lack of understanding for the physical model, then one can resort to the procedure of plotting the data on all types of probability paper to determine which best describes the experiment. Interpretations of the results of this approach should be handled carefully. It is quite possible that a model may describe the system within the ranges of data available but beyond this point may be in serious error. As in all statistical calculations and procedures, it is imperative that judgment be used in applying the results of the test.

Probability plotting provides a pictorial representation of the data, an estimate of the goodness of the fit to the probability model, estimates of the percentiles of the distribution, and estimates of the distribution parameters.

These estimates can be obtained from censured data or when only k of n observations are known. Frequently experiments are terminated when it might be desirable to have more data for statistical analysis, but it is necessary to estimate the various statistical parameters. Probability plotting provides an easy way to obtain these estimates.

Procedures:

a. Select the proper probability paper designed for the distribution to be considered. Probability paper is available at all engineering and mathematical supply houses.
b. Rank the observations from the smallest to the largest in magnitude.
c. Plot the x_i's on the probability paper versus $(i - 1/2)100/n$. If the probability paper is marked according to the proportion of observations, eliminate the 100 in the above equation. The following examples will be based upon percent observations.
d. Examine the plot and draw a straight line by eye on the graph. If the chosen model or probability paper is correct, the points will cluster around the line, but there will be some deviation because of random sampling fluctuations. Judgment must be exercised in deciding what constitutes a straight line. Obviously different observers will reach different conclusions. When the data obviously deviate from the straight line, it may be necessary to resort to computational statistics to determine the reliability of the model to describe the data.
e. Estimating parameters from a probability plot. In a normal distribution the mean and median coincide; therefore, the value plotted on the

probability paper versus the 50 percentile figure represents both the mean and median. The standard deviation can be estimated by using the fact that for any normal distribution the standard deviation equals approximately two fifths of the difference between the 90th and the 10th percentiles. A probability plot may also be used to estimate the probability that a specified value will be exceeded. This procedure is illustrated in the solution to the following problem.

Results and Computations:

The steps outlined under Method and Materials were followed and each step is identified by the letters used in the above referenced section.

a. The objective of the exercise is to evaluate if the data fit a normal distribution; therefore, normal probability paper was selected.

b. & c.   Ranked Data and Calculation of Percentage Observations

Rank	Effluent COD (A)	$\% \geqslant \dfrac{(i - \frac{1}{2})100}{n}$	Effluent Filtrate COD (B)
1	17	2.78	8
2	21	8.33	13
3	21	13.89	15
4	21	19.44	18
5	25	24.99	21
6	28	30.56	22
7	31	36.11	23
8	33	41.67	24
9	35	47.22	25
10	35	52.78	26
11	36	58.33	26
12	41	63.89	28
13	49	69.44	28
14	52	75.00	29
15	56	80.55	29
16	57	86.11	30
17	63	91.67	36
18	65	97.22	36

$$\Sigma_{A_j} = 686 \qquad\qquad \Sigma_{B_j} = 437$$

$$\bar{x}_A = 38.1 \qquad\qquad \bar{x}_B = 24.3$$

d. See Figure 9-1.

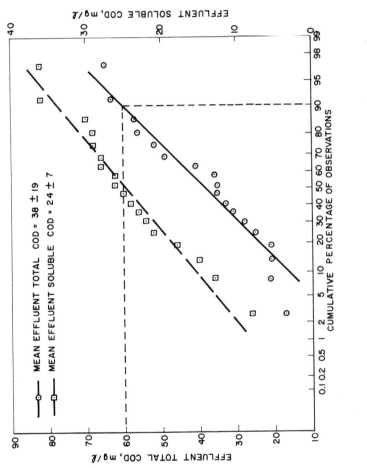

Figure 9-1. Normal probability plot to determine percent distribution and distribution parameters.

e. Estimation of Standard Deviation

St. Dev. = 2/5 (90% - 10%)

Mean Total COD = 2/5 (60 - 13) = 2/5 (47) = 18.8 or 19

Mean Soluble COD = 2/5 (33.5 - 15.5) - 2/5 (18.0)
                  = 7.2 or 7

Probability That Value Will Be Exceeded:

Locate the value of 60 mg/l on the probability plot (Figure 9-1). Find the intersection of the line drawn on the plot and the 60 mg/l value. Read the corresponding value on the percentage scale, or 90 percentile. This indicates that there is a 0.10 probability that this effluent concentration will be exceeded.

Conclusions:

Both sets of effluent quality data are described by a normal distribution and the probability plot is useful in estimating the statistical parameters as shown by the standard statistical tests.

Exercise Number 10

Nonlinear Regression and Correlation

Problem:

1. Determine the regression coefficient by regression of log Y upon log X.
2. Test $b_{\log y \,.\, \log x}$ for significance by means of F and t tests. Show that the two tests are identical.
3. Calculate the standard error of estimate.
4. Determine the regression equation and use it to prepare the regression line.
5. Determine the 95% confidence limits for the regression line. Enter the latter on the regression chart.
6. Determine the correlation coefficient.
7. Test it for significance by the F and t tests and show that these give identical conclusions as the corresponding tests for the regression coefficient.
8. Interpret data.

A 4" schedule 40 steel pipe flowing full of water, Y = velocity in feet per second, and X = pressure drop per 100 feet, yields the following relationship between Y and X:

X, psi/100 ft	Y, ft/sec
2.02	0.180
3.30	0.415
4.10	0.774
5.80	1.23
6.93	1.79
8.50	2.47
9.10	3.25
10.50	4.12
12.60	5.65

Purpose:

To determine if a significant log-log linear regression exists between the velocity of water in a pipe and the pressure loss per length of pipe.

Method and Materials:

A linear regression analysis of the logarithms of the variables was used to evaluate the data.

Results and Computations:

Data:

X	log X	Y	log Y	(log X)(log Y)	(log X)^2	(log Y)^2
0.180	9.25527-10	2.02	0.30535	9.77260-10	0.55462	0.09324
0.415	9.61805-10	3.30	0.51851	9.80196-10	0.14589	0.26885
0.774	9.88874-10	4.10	0.61278	9.93182-10	0.01238	0.37550
1.23	0.08991	5.80	0.76343	0.06864	0.00808	0.58283
1.79	0.25285	6.93	0.84073	0.21258	0.06393	0.70683
2.47	0.39270	8.50	0.92942	0.36498	0.15421	0.86382
3.25	0.51188	9.10	0.95904	0.49091	0.26202	0.91982
4.12	0.61490	10.50	1.02119	0.62793	0.37810	1.04283
5.65	0.75205	12.60	1.10037	0.82753	0.56558	1.21081
	1.37635		7.05082	2.09895	2.14481	6.06453

$$\log \bar{x} = 0.15293 \qquad \log \bar{y} = 0.78342$$
$$\bar{x} = 1.42 \qquad\qquad \bar{y} = 6.07$$

$$\Sigma(\log y)^2 = \Sigma(\log Y)^2 - \frac{(\Sigma \log Y)^2}{n} = 6.06453 - \frac{(7.05082)^2}{9}$$
$$= 0.54075$$

$$\Sigma(\log x)^2 = \Sigma(\log X)^2 - \frac{(\Sigma \log X)^2}{n} = 2.14481 - \frac{(1.37635)^2}{9}$$
$$= 1.93433$$

$$\Sigma(\log x)(\log y) = \Sigma(\log X)(\log Y) - \frac{\Sigma(\log X)\, \Sigma(\log Y)}{n}$$
$$= 2.09895 - \frac{(1.37635)(7.05082)}{9}$$
$$= 1.02068$$

Prior to any calculations the data were plotted on log-log graph paper to determine if the relationship approximated a straight line. This is shown in Figure 10-1 along with the final results which were calculated as follows.

Testing Significance of Regression:

Analysis of Variance Table

Source	d.f.	SS	MS	F
Regression	1	$(1.02068)^2\,1.93433 = 0.53858$	0.53858	1,737.35**
Residual	7	0.00217	0.00031	
	8	0.54075		

The F-value is highly significant which indicates that log Y is dependent upon the value of log X, or that a linear relationship exists.

Prediction Equation:

$$\log \hat{Y} = \log \bar{y} + b_1\,(\log X - \log \bar{x})$$
$$\log \hat{Y} = b_0 + b_1\,\log X \text{ where } b_0 = \log \bar{y} - b\,\log \bar{x}$$
$$b_1 = \frac{\Sigma(\log x \log y)}{\Sigma(\log x)^2} = \frac{1.02068}{1.93433} = \underline{\underline{0.52767}}$$
$$b_0 = 0.78342 - 0.52767\,(0.15293) = \underline{\underline{0.70272}}$$
$$\underline{\underline{\log \hat{Y} = 0.7027 + 0.5277 \log X}}$$

Computation of Points Used to Sketch the Prediction Equation:

When X = 1.0, log X = 0
$$\log \hat{Y} = 0.7027 + 0.5277(0) = \underline{\underline{0.7027}}$$
$$\hat{Y} = \underline{\underline{5.04}} \text{ ft/sec}$$

When X = 10.0, log X = 1.000
$$\log \hat{Y} = 0.7027 + 0.5277(1.000) = \underline{\underline{1.2304}}$$
$$\hat{Y} = \underline{\underline{17.0}} \text{ ft/sec}$$

The predicted values were plotted on Figure 10-1 and joined by a straight line. The intercept value (b_0) represents the point at which the regression line crosses the value of 1.0 on the abscissa.

Determination of 95% Confidence Limits for Regression Line:

$$\log \hat{Y} = \log \bar{y} + b_1 \log X$$

The variance of $\log \hat{Y}$ is equal to:

$$s^2_{\log \hat{Y}} = \frac{s_{\log y \log x}^2}{n} + \frac{s_{\log y \cdot \log x}^2}{\Sigma (\log x)^2} (\log x)^2$$

$$s_{\log \hat{Y}} = \sqrt{s_{\log y \cdot \log x}^2 \left\{ \frac{1}{n} + \frac{(\log x)^2}{\Sigma (\log X)^2} \right\}}$$

where $\log x = \log X - \log \bar{x}$

$$s_{\log \hat{Y}} = \sqrt{0.00031 \left\{ \frac{1}{9} + \frac{(\log x)^2}{1.93433} \right\}}$$

95% C.L. for $\log \hat{Y}$ are: $\log \hat{Y} \pm t_{0.05(n-2)} \cdot s_{\log \hat{Y}}$

95% C.L. for $\log \hat{Y}$ at $\bar{x}$ are:

$$\log \hat{Y} = \log \bar{y}$$

$$\log \bar{y} \pm t_{.05} \cdot s_{\log \hat{Y}} = 0.78342 \pm 2.365 \sqrt{0.0000344}$$

$$= 0.78342 \pm 0.01387$$

95% C.L. = 0.79729 to 0.76955 or 6.27 to 5.88

95% C.L. For $\log \hat{Y}$ at Values Other Than $\bar{x}$:

Let $X = 10, \log X = 1.000$

$$s_{\log \hat{Y}} = \sqrt{0.00031 \left\{ \frac{1}{9} + \frac{(1.000 - 0.15293)^2}{1.93433} \right\}} = 0.01222$$

$$\log \hat{Y} \pm t_{.05} s_{\log \hat{Y}} = 1.2304 \pm 2.365 (0.01222)$$

$$= 1.2304 \pm 0.0289$$

95% C.L. = 1.2593 to 1.2015 or 18.17 to 15.90

Let $X = 0.2, \log X = 9.30103 - 10$

$$s_{\log \hat{Y}} = \sqrt{0.00031 \left\{ \frac{1}{9} + \frac{(9.30103 - 10 - 0.15293)^2}{1.93433} \right\}}$$

$$s_{\log \hat{Y}} = 0.01228$$

$$\log \hat{Y} \pm t_{.05} \, s_{\log \hat{Y}} = 0.33385 \pm 2.365 \, (0.01228)$$

$$= 0.33385 \pm 0.02904$$

95% C.L. $= 0.36289$ to 0.30481 or 2.31 to 2.02

Let $X = 4.0, \log X = 0.60206$

$$s_{\log \hat{Y}} = \sqrt{0.00031 \left\{ \frac{1}{9} + \frac{(0.60206 - 0.15293)^2}{1.93433} \right\}} = \underline{0.00910}$$

$$\log \hat{Y} \pm t_{.05} \, s_{\log \hat{Y}} = 1.02040 \pm 2.365 \, (0.00910)$$

$$= 1.02040 \pm 0.02152$$

95% C.L. $= 1.04192$ to 0.99888 or 11.01 to 9.97

The 95% C.L. values calculated above were then plotted on Figure 10-1 and joined by a continuous line.

Testing Regression Coefficient by the t-test:

$$t = \frac{b_1 - \beta_1}{s_{b_1}}$$

$$H_0 : \beta_1 = 0$$

$$t = \frac{b_1}{s_{b_1}}$$

$$t_{0.05 \atop 0.01}{}^{(n-2)} = 2.365$$
$$= 3.499$$

$$s_{b_1} = \sqrt{\frac{s_{\log y \cdot \log x}^2}{\Sigma (\log x)^2}} = \sqrt{\frac{s_e^2}{\Sigma (\log x)^2}} = \sqrt{\frac{0.00031}{1.93433}}$$

$$s_{b_1} = \underline{0.01266}$$

$$t = \frac{0.52767}{0.01266} = \underline{41.68009 **}$$

$$t^2 = F$$

$F = (41.68009)^2 = 1737.23$ which is approximately equal to 1737.35. This is within rounding error.

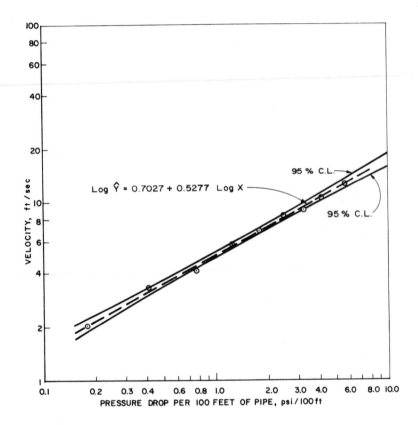

Figure 10-1. Plot of data showing results of log-log regression and correlation.

Determination of Correlation Coefficient:

$$r = \frac{\Sigma(\log x)(\log y)}{\sqrt{\Sigma(\log y)^2\,\Sigma(\log x)^2}} = \frac{1.02068}{\sqrt{(0.54075)(1.93433)}}$$

$$r = \underline{0.99799}$$

Testing Significance of Correlation Coefficient:

To show that conclusions will be identical with tests for the regression coefficient.

Analysis of Variance Table

Source	d.f.	SS	MS	F
Regression	1	$r^2\,\Sigma(\log y)^2$	$r^2\,\Sigma(\log y)^2/1$	MSR/MSE
Error	n-2	$(1\text{-}r^2)\Sigma(\log y)^2$	$(1\text{-}r^2)\Sigma(\log y)^2/n\text{-}2$	
Total	n-1	$\Sigma(\log y)^2$		

$$F = \frac{\dfrac{r^2\,\Sigma(\log y)^2}{1}}{\dfrac{(1\text{-}r^2)\,\Sigma(\log y)^2}{n\text{-}2}} = \frac{r^2\,\Sigma(\log y)^2\,(n\text{-}2)}{(1\text{-}r^2)\,\Sigma(\log y)^2} = \frac{r^2\,(n\text{-}2)}{(1\text{-}r^2)}$$

$$t = \frac{r\,\sqrt{n\text{-}2}}{\sqrt{1\text{-}r^2}}$$

For 1 d.f., $t = \sqrt{F}$

Analysis of Variance

Source	d.f.	SS	MS	F
Regression	1	0.53858	0.53858	1737.35**
Error	7	0.00217	0.00031	
Total	8			

F values for the regression coefficient and the correlation coefficient are in agreement. It was necessary to carry 10 significant figures to obtain such close agreement. Slight deviations in rounding techniques can cause significant rounding errors.

$$t = \frac{0.99799 \sqrt{9 \cdot 2}}{\sqrt{1 - (0.99799)^2}} = \frac{2.64043}{0.063372} = \underline{\underline{41.67}}**$$

which also agrees within rounding errors:

$$41.67 \approx 41.68$$

Conclusions:

There is a definite log-log linear relationship between the velocity of the water in the pipe and the pressure drop per length of pipe. All tests indicate a highly significant (95% C.L.) relationship, and an examination of the plotted data on log-log graph paper also indicates that the statistical tests should verify the log-log linear relationship.

Exercise Number 11

Multiple and Partial Regression and Correlation

Problem:

1. Determine the partial regression coefficients by regression of Y upon X_1, X_2, and X_3.
2. Test the partial regression coefficients, b_1, b_2, and b_3 for significance by the F and t tests.
3. Test the significance of the multiple regression.
4. Calculate the standard error of the estimate and the standard errors for the partial regression coefficients.
5. Determine the multiple correlation coefficient.
6. Calculate the standard partial regression coefficients.
7. Delete one independent variable and recalculate the multiple regression equation.
8. Interpret the results.

A study of the accumulation of sludge in municipal sewage lagoons produced the results shown in Table 11-1.

Purpose:

To determine if a linear multiple regression analysis describes the relationship between sludge accumulation and the three independent variables, age of the lagoon, BOD_5 applied in pounds per acre per day, and the total solids found in the deposited solids.

Table 11-1. Summary of the Data for the Lagoons Studies[a]

Location of Lagoon	Surface Area Acres	Age of Lagoon at Sampling Date Months	B.O.D. Applied Lbs. Per Acre Per Day	Composited Sample Total Solids %	Average Sludge Depth in Inches
Brandon-W	5.00	82.0	35.0	3.48	6.86
Brandon-E	5.00	82.0	18.0	10.30	3.92
Fayette-N	4.19	65.0	23.0	17.40	4.90
Fayette-SW	3.09	65.0	23.0	19.20	4.11
Newton	3.00	54.0	28.0	4.92	4.68
Rolling Fork-E	5.13	54.0	25.0	5.58	4.84
Rolling Fork-W	5.07	54.0	25.0	5.62	4.85
Ellisville	3.50	44.0	28.5	11.30	2.61
Canton Club	3.00	31.0	70.0	11.90	5.54
Canton	5.00	26.0	15.0	7.04	2.34
Utica	6.00	25.0	7.5	13.40	3.29
Decatur	3.00	25.0	10.5	4.34	2.29
Walnut Grove	2.00	19.0	25.0	3.66	3.56
Ridgeland-N	3.00	5.0	8.4	9.70	1.03
Philadelphia	3.00	5.0	7.5	10.90	0.16

[a]Taken from Middlebrooks, E. J., A. J. Panagiotou, and H. K. Williford; "Sludge Accumulation in Municipal Sewage Lagoons," Water and Sewage Works, February 1965.

Methods and Materials:

A linear multiple regression and correlation analysis was employed to evaluate the data.

Results and Computations:

Data:

Sample Number	Age of Lagoon X_1	BOD$_5$ Applied X_2	Sludge, % Total Solids X_3	Sludge Depth Y
1	82.0	35.0	3.48	6.86
2	82.0	18.0	10.30	3.92
3	65.0	23.0	17.40	4.90
4	65.0	23.0	19.20	4.11
5	54.0	28.0	4.92	4.68

Sample Number	Age of Lagoon X_1	BOD$_5$ Applied X_2	Sludge, % Total Solids X_3	Sludge Depth Y
6	54.0	25.0	5.58	4.84
7	54.0	25.0	5.62	4.85
8	44.0	28.5	11.30	2.61
9	31.0	70.0	11.90	5.54
10	26.0	15.0	7.04	2.34
11	25.0	7.5	13.40	3.29
12	25.0	10.5	4.34	2.29
13	19.0	25.0	3.66	3.56
14	5.0	8.4	9.70	1.03
15	5.0	7.5	10.90	0.16
ΣX_i	636.0	349.40	138.74	54.98
$\bar{x}$	42.40	23.29	9.25	3.67
ΣX_i^2	35,880.0	11,496.6	1,620.1	244.99

$\Sigma X_i X_j$ $\Sigma X_1 X_2 = 16,366.5$ $\Sigma X_1 X_3 = 6,044.6$ $\Sigma X_1 Y = 2,806.1$

$\Sigma X_2 X_3 = 3,228.2$ $\Sigma X_2 Y = 1,536.0$

$\Sigma X_3 Y = 496.4$

Table 11-2. Sum of Products and Simple Correlations

	X_1	X_2	X_3	Y
X_1	$\Sigma x_1^2 = 8,913.6$	$\Sigma x_1 x_2 = 1,551.9$ $r_{12} = 0.2837$	$\Sigma x_1 x_3 = 162.0$ $r_{13} = 0.0935$	$\Sigma x_1 y = 474.9$ $r_{y1} = 0.7630$
X_2		$\Sigma x_2^2 = 3,357.9$	$\Sigma x_2 x_3 = -3.52$ $r_{23} = 0.0033$	$\Sigma x_2 y = 255.3$ $r_{y2} = 0.6684$
X_3			$\Sigma x_3^2 = 336.8$	$\Sigma x_3 y = -12.13$ $r_{y3} = 0.1003$
Y				$\Sigma y^2 = 43.5$

Normal Equations:

$$x_1: \quad b_1 \Sigma x_1^2 + b_2 \Sigma x_1 x_2 + b_3 x_1 x_3 = \Sigma x_1 y$$

$$x_2: \quad b_1 \Sigma x_1 x_2 + b_2 \Sigma x_2^2 + b_3 \Sigma x_3 x_2 = \Sigma x_2 y$$

$$x_3: \quad b_1 \Sigma x_1 x_3 + b_2 \Sigma x_2 x_3 + b_3 \Sigma x_3^2 = \Sigma x_3 y$$

Multiple Regression Equation:

$$\hat{Y} = a + b_1 X_1 + b_2 X_2 + b_3 X_3$$

$$\hat{Y} = \bar{y} + b_1 x_1 + b_2 x_2 + b_3 x_3$$

Where $x_1 = (X_1 - \bar{x}_1)$, etc.

Substitute Values into Normal Equations:

$$8{,}913.6\, b_1 + 1{,}551.9\, b_2 + 162.0\, b_3 = 474.9$$

$$1{,}551.9\, b_1 + 3{,}357.9\, b_2 - 3.52\, b_3 = 255.3$$

$$162.0\, b_1 - 3.52\, b_2 + 336.8\, b_3 = -12.13$$

Solve Normal Equations:

Cramer (1704-1752) showed that the values of b_1, b_2, and b_3 in equations such as the three normal equations shown above can be obtained using determinants as follows:

$$b_1 = \frac{\begin{vmatrix} \Sigma x_1 y & \Sigma x_1 x_2 & \Sigma x_1 x_3 \\ \Sigma x_2 y & \Sigma x_2^2 & \Sigma x_2 x_3 \\ \Sigma x_3 y & \Sigma x_2 x_3 & \Sigma x_3^2 \end{vmatrix}}{\begin{vmatrix} \Sigma x_1^2 & \Sigma x_1 x_2 & \Sigma x_1 x_3 \\ \Sigma x_1 x_2 & \Sigma x_2^2 & \Sigma x_2 x_3 \\ \Sigma x_1 x_3 & \Sigma x_2 x_3 & \Sigma x_3^2 \end{vmatrix}} = \frac{D_1}{D}$$

$$b_2 = \frac{\begin{vmatrix} \Sigma x_1^2 & \Sigma x_1 y & \Sigma x_1 x_3 \\ \Sigma x_1 x_2 & \Sigma x_2 y & \Sigma x_2 x_3 \\ \Sigma x_1 x_3 & \Sigma x_2 x_3 & \Sigma x_3^2 \end{vmatrix}}{D} = \frac{D_2}{D}$$

$$b_3 = \frac{\begin{vmatrix} \Sigma x_1^2 & \Sigma x_1 x_2 & \Sigma x_1 y \\ \Sigma x_1 x_2 & \Sigma x_2^2 & \Sigma x_2 y \\ \Sigma x_1 x_3 & \Sigma x_2 x_3 & \Sigma x_3 y \end{vmatrix}}{D} = \frac{D_3}{D}$$

Substitute Values into Determinants:

$$D = \begin{vmatrix} 8,913.6 & 1,551.9 & 162.0 \\ 1,551.9 & 3,357.9 & -3.52 \\ 162.0 & -3.52 & 336.8 \end{vmatrix}$$

$$D_1 = \begin{vmatrix} 474.9 & 1,551.9 & 162.0 \\ 255.3 & 3,357.9 & -3.52 \\ -12.13 & -3.52 & 336.8 \end{vmatrix}$$

$$D_2 = \begin{vmatrix} 8,913.6 & 474.9 & 162.0 \\ 1,551.9 & 255.3 & -3.52 \\ 162.0 & -12.13 & 336.8 \end{vmatrix}$$

$$D_3 = \begin{vmatrix} 8,913.6 & 1,551.9 & 474.9 \\ 1,551.9 & 3,357.9 & 255.3 \\ 162.0 & -3.52 & -12.13 \end{vmatrix}$$

Expansion of Determinants:

$$D_i = \begin{vmatrix} a_{11} & a_{12} & a_{13} \\ a_{21} & a_{22} & a_{23} \\ a_{31} & a_{32} & a_{33} \end{vmatrix} = a_{11} \begin{vmatrix} a_{22} & a_{23} \\ a_{32} & a_{33} \end{vmatrix} - a_{12} \begin{vmatrix} a_{21} & a_{23} \\ a_{31} & a_{33} \end{vmatrix} + a_{13} \begin{vmatrix} a_{21} & a_{22} \\ a_{31} & a_{32} \end{vmatrix}$$

Solution to 2-rowed Determinants:

$$\begin{vmatrix} a_{11} & a_{12} \\ a_{21} & a_{22} \end{vmatrix} = a_{11}a_{22} - a_{12}a_{21}$$

$$D = 8,913.6 \begin{vmatrix} 3,357.9 & -3.52 \\ -3.52 & 336.8 \end{vmatrix} - 1,551.9 \begin{vmatrix} 1,551.9 & -3.52 \\ 162.0 & 336.8 \end{vmatrix}$$

$$+ 162.0 \begin{vmatrix} 1,551.9 & 3,357.9 \\ 162.0 & -3.52 \end{vmatrix}$$

$$D = 8,913.6 \; [1,130,941 + 12.39] - 1.551.9 \; [522,680 - 570.2]$$
$$+ 162.0 \; [-5462.7 - 543,980]$$

$$D = 100.80866 \times 10^8 - 8.102622 \times 10^8 - 0.8900972 \times 10^8$$

$$D = \underline{\underline{91.816 \times 10^8}}$$

$$D_1 = 474.9 \begin{vmatrix} 3,357.9 & -3.52 \\ -3.52 & 336.8 \end{vmatrix} - 1,551.9 \begin{vmatrix} 255.3 & -3.52 \\ -12.13 & 336.8 \end{vmatrix}$$

$$+ 162 \begin{vmatrix} 255.3 & 3,357.9 \\ -12.13 & -3.52 \end{vmatrix}$$

$$D_1 = 474.9 \; (1,130,940.7 - 12.4) - 1,551.9 \; (85,985.0 - 42.7)$$
$$+ 162 \; (-898.7 + 40,731.3)$$

$$D_1 = 5.3707785 \times 10^8 - 1.333739 \times 10^8 + 0.0645 \times 10^8$$

$$D_1 = \underline{\underline{4.10154 \times 10^8}}$$

$$D_2 = 8,913.6 \begin{vmatrix} 255.3 & -3.52 \\ -12.13 & 336.8 \end{vmatrix} -474.9 \begin{vmatrix} 1,551.9 & -3.52 \\ 162.0 & 336.8 \end{vmatrix}$$

$$+ 162 \begin{vmatrix} 1,551.9 & 255.3 \\ 162.0 & -12.13 \end{vmatrix}$$

D_2 = 8,913.6 (8.5985 x 10^4 - 42.6976) - 474.9 (5.2268 x 10^5
$\qquad$ + 5.7024 x 10^2) + 162 (-1.8825 x 10^4 - 4.13586 x 10^4)

D_2 = 7.6606 x 10^8 - 2.4849 x 10^8 - 0.09749743 x 10^8

D_2 = 5.0782 x 10^8

$$D_3 = 8,913.6 \begin{vmatrix} 3,357.9 & 255.3 \\ -3.52 & -12.13 \end{vmatrix} -1,551.9 \begin{vmatrix} 1,551.9 & 255.3 \\ 162.0 & -12.13 \end{vmatrix}$$

$$+ 474.9 \begin{vmatrix} 1,551.9 & 3,357.9 \\ 162.0 & -3.52 \end{vmatrix}$$

D_3 = 8.913.6 (-4.07313 x 10^4 + 898.656) - 1,551.9 (-1.8825 x 10^4
$\qquad$ -4.13586 x 10^4) + 474.9 (-5462.69 - 5.43980 x 10^5)

D_3 = -3.5505 x 10^8 + 0.934 x 10^8 - 2.6093 x 10^8 = 5.2258 x 10^8

$$b_1 = \frac{D_1}{D} = \frac{4.10154 \times 10^8}{91.816 \times 10^8} = \underline{\underline{0.04467}}$$

$$b_2 = \frac{D_2}{D} = \frac{5.0782 \times 10^8}{91.816 \times 10^8} = \underline{\underline{0.055308}}$$

$$b_3 = \frac{D_3}{D} = \frac{5.2258 \times 10^8}{91.816 \times 10^8} = \underline{\underline{-0.056916}}$$

Multiple Regression Equation:

$\hat{Y} = \bar{y} + b_1 x_1 + b_2 x_2 + b_3 x_3$
$\hat{Y} = 3.67 + 0.04467 (X_1 - 42.40) + 0.05531 (X_2 - 23.29)$
$\qquad - 0.05692 (X_3 - 9.25)$
$\hat{Y} = 2.656 + 0.04467 X_1 + 0.05531 X_2 - 0.05692 X_3$

Check: $b_1 \Sigma x_1{}^2 + b_2 \Sigma x_1 x_2 + b_3 \Sigma x_1 x_3 = \Sigma x_1 y$

$\qquad$ 0.04467 (8,913.6) + 0.055308 (1,551.9) - 0.056916(162.0) = $\Sigma x_1 y$

$\qquad$ 398.17 + 85.84 - 9.22 = 474.8   which is within rounding error of
474.9, the observed value of $\Sigma x_1 y$.

$\qquad$ Regression SS = $b_1 \Sigma x_1 y + b_2 \Sigma x_2 y + b_3 \Sigma x_3 y$

$$RSS = 0.04467(474.9) + 0.05531(255.3) - 0.05692(-12.13)$$
$$RSS = 21.214 + 14.121 + 0.690 = \underline{\underline{36.025}}$$

Test of Significance of Multiple Regression:

The reduction in sum of squares attributable to regression can be evaluated by the F-value (Table 11-3).

Table 11-3. Analysis of Variance to Test Significance of Regression

Source	Symbolic		Data From Table 11-2			
	d.f.	SS	d.f.	SS	MS	F
Regression on k variables	k	$\sum_i b_i (\sum x_1 y)$	3	36.025	12.01	17.66**
Residual	n-k-1	By substraction	11	7.48	0.68	
Total	n-1	$\sum y^2$	14	43.5		

Tabular value of F from Table A-2

$$F_{0.05} = 3.59$$
$$F_{0.01} = 6.22$$

Multiple Correlation Coefficient:

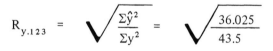

$$R_{y.123} = \sqrt{\frac{\sum \hat{y}^2}{\sum y^2}} = \sqrt{\frac{36.025}{43.5}}$$

$$R_{y.123} = \underline{\underline{0.910^{**}}}$$

Significance levels for R from Table A-5

$$R_{0.05} = 0.703$$
$$R_{0.01} = 0.793$$

$R_{y.123}$ measures the combined effect of the independent variables X_1, X_2, and X_3 on the dependent variable Y. The numerical value of the multiple correlation coefficient is always at least equivalent to the simple or partial correlation.

Standard Errors and Tests of Significance for Partial Regression Coefficients:

The standard error of the estimate of the populations of Y values is:

$$s_{y.1...k} = \sqrt{\frac{\Sigma(Y - \hat{Y})^2}{n-k-1}}$$

$$\Sigma(Y - \hat{Y})^2 = \Sigma y^2 - \underset{i}{\Sigma} b_i (\Sigma x_i y)$$

$$s_{y.123} = \sqrt{\frac{43.5 - 36.025}{11}} = \underline{0.8243} \text{ inches of sludge depth}$$

To obtain standard errors for the partial regression coefficients it is necessary to obtain the inverse of the determinant D. The inverse of D can be obtained as follows:

$$D = 9.1816 \times 10^9 = \begin{vmatrix} a_{11} & a_{12} & a_{13} \\ a_{21} & a_{22} & a_{23} \\ a_{31} & a_{32} & a_{33} \end{vmatrix}$$

$$D^{-1} = \begin{vmatrix} a_{11} & a_{12} & a_{13} \\ a_{21} & a_{22} & a_{23} \\ a_{31} & a_{32} & a_{33} \end{vmatrix}$$

$$a_{11} = (a_{22}a_{33} - a_{32}a_{23})/D$$

$$a_{21} = (a_{31}a_{23} - a_{21}a_{33})/D$$

$$a_{31} = (a_{21}a_{32} - a_{31}a_{22})/D$$

$$a_{12} = (a_{32}a_{13} - a_{12}a_{33})/D$$

$$a_{22} = (a_{11}a_{33} - a_{31}a_{13})/D$$

$$a_{32} = (a_{31}a_{12} - a_{11}a_{32})/D$$

$$a_{13} = (a_{12}a_{23} - a_{22}a_{13})/D$$

$$a_{23} = (a_{21}a_{13} - a_{11}a_{23})/D$$

$$a_{33} = (a_{11}a_{22} - a_{21}a_{12})/D$$

$$a_{11} = [(3,357.9)(336.8) - (-3.52)(-3.52)]/9.1816 \times 10^9$$

$$a_{11} = \underline{0.00012317}$$

$$a_{21} = [(162.0)(-3.52) - (1,551.9)(336.8)]/D$$

$$a_{21} = \underline{-0.000056989}$$

$$a_{31} = [(1,551.9)(-3.52) - (162.0)(3,357.9)]/D$$

$$a_{31} = \underline{-0.000059842}$$

$$a_{12} = [(-3.52)(162.0) - (1,551.9)(336.8)]/D$$

$$a_{12} = \underline{\underline{-0.000056989}}$$

$$a_{22} = [(8,913.6)(336.8) - (162.0)(162.0)]/D$$

$$a_{22} = \underline{\underline{0.00032411}}$$

$$a_{32} = [(162.0)(1,551.9) - (8,913.6)(-3.52)]/D$$

$$a_{32} = \underline{\underline{0.000030799}}$$

$$a_{13} = [(1,551.9)(-3.52) - (3,357.9)(162.0)]/D$$

$$a_{13} = \underline{\underline{-0.000059842}}$$

$$a_{23} = [(1,551.9)(162) - (8,913.6)(-3.52)]/D$$

$$a_{23} = \underline{\underline{0.000030799}}$$

$$a_{33} = [(8,913.6)(3,357.9) - (1,551.9)(1,551.9)]/D$$

$$a_{33} = \underline{\underline{0.0029976}}$$

Partial regression coefficients also can be obtained using the inverse matrix.

$$b_1 = a_{11}\Sigma x_1 y + a_{12}\Sigma x_2 y + a_{13}\Sigma x_3 y$$

$$b_1 = (0.00012317)(474.9) + (-0.000056989)(255.3)$$
$$+ (-0.000059842)(-12.13)$$

$$b_1 = \underline{\underline{0.04467}} \quad \text{which agrees with earlier value.}$$

$$b_2 = a_{21}\Sigma x_1 y + a_{22}\Sigma x_2 y + a_{23}\Sigma x_3 y$$

$$b_3 = a_{31}\Sigma x_1 y + a_{32}\Sigma x_2 y + a_{33}\Sigma x_3 y$$

A general expression for the standard errors of the partial regression coefficients is:

$$s_{b_i} = \sqrt{a_{ii}\, s_{y.123}^2}$$

$$s_{b_1} = \sqrt{a_{11}\, s_{y.123}^2} = \sqrt{0.00012317\,(0.8243)}$$

$$s_{b_1} = \underline{\underline{0.010076}}$$

$$s_{b_2} = \sqrt{a_{22}\, s_{y.123}^2} = \sqrt{0.00032411\,(0.8243)}$$

$$s_{b_2} = \underline{\underline{0.016345}}$$

$$s_{b_3} = \sqrt{a_{33}\, s_{y.123}^2} = \sqrt{0.0029976\,(0.8243)}$$

$$s_{b_3} = \underline{\underline{0.049708}}$$

The significance of partial regression coefficients is tested with the t test.

$$H_o: \beta_i = \beta_{io} = 0$$

$$t = \frac{b_i - \beta_{io}}{s_{b_i}}, \quad \text{d.f.} = n\text{-}k\text{-}1$$

$$t = \frac{b_i}{s_{b_i}}$$

$$t = \frac{b_1}{s_{b_1}} = \frac{0.04467}{0.010076} = \underline{\underline{4.433**}}$$

$$t = \frac{b_2}{s_{b_2}} = \frac{0.055308}{0.016345} = \underline{\underline{3.384**}}$$

$$t = \frac{b_3}{s_{b_3}} = \frac{-0.056916}{0.049708} = \underline{\underline{-1.145^{n.s.}}}$$

Significance levels for t from Table A-1.

$$\text{d.f} = 15 - 3 - 1 = 11$$

$$t_{0.05} = 2.201$$
$$0.01 = 3.106$$

The contribution to the regression equation by X_3 is not significant and a better estimate of Y would be obtained without X_3, or other variables should be introduced. Partial correlation coefficients can be calculated but the results will yield the same conclusions reached with the significance tests of the partial regression coefficients.

Standard Partial Regression Coefficients:

Standard partial regression coefficients are independent of the original units of measurement; therefore, a comparison of the coefficients indicates the relative importance of the independent variables. If b_1' is three times as large as b_2', then X_1 is approximately three times as important as X_2 in estimating values of Y.

$$b_i' = b_i \frac{s_i}{s_y} = b_i \frac{\sqrt{\dfrac{\Sigma x_i^2}{d.f.}}}{\sqrt{\dfrac{\Sigma y^2}{d.f.}}}$$

$$b_1' = b_1 \frac{s_1}{s_y} = 0.04467 \frac{\sqrt{8,913.6/14}}{\sqrt{43.5/14}}$$

$$b_1' = 0.04467 \left(\frac{25.23}{1.763}\right) = \underline{\underline{0.63926}}$$

$$b_2' = b_2 \frac{s_2}{s_y} = 0.055308 \frac{\sqrt{3,357.9/14}}{\sqrt{43.5/14}}$$

$$b_2' = 0.055308 \left(\frac{15.487}{1.763}\right) = \underline{\underline{0.48585}}$$

$$b_3' = b_3 \frac{s_3}{s_y} = -0.056916 \frac{\sqrt{336.8/14}}{\sqrt{43.5/14}}$$

$$b_3' = -0.056916 \left(\frac{4.905}{1.763}\right) = \underline{\underline{0.15835}}$$

From the values of b_1', b_2', and b_3', it can be seen that X_1 is approximately 1.3 times as useful as X_2 and four times as useful as X_3 in estimating Y; whereas, X_2 is approximately three times as useful as X_3 in estimating Y. Therefore, X_1 and X_2 are much more influential in estimating values of Y, and X_3 could probably be eliminated from the multiple regression equation without losing much precision.

Deletion of Independent Variable:

When the results of a multiple regression analysis indicate that certain variables are not significant, it is best to delete this variable and calculate another multiple regression equation. Obviously, one can start with new normal equations, but in most cases it is easier to use the following procedure.

$$a_{ij}' = a_{ij} - \frac{a_{ik} a_{jk}}{a_{kk}}$$

a_{ij}' is the value that would be obtained if X_k had been excluded in the original calculations.

$$a_{11}' = a_{11} - \frac{a_{13}^2}{a_{33}} = 0.00012317 - \frac{(-0.000059842)^2}{0.0029976}$$

$$a_{11}' = \underline{0.00012198}$$

$$a_{12}' = a_{12} - \frac{a_{13}a_{23}}{a_{33}} = -0.000056989 - \frac{(-0.000059842)(0.000030799)}{0.0029976}$$

$$a_{12}' = -0.000056374$$

$$a_{22}' = a_{22} - \frac{a_{23}^2}{a_{33}} = 0.00032411 - \frac{(0.000030799)^2}{0.0029976}$$

$$a_{22}' = \underline{0.00032379}$$

Check: $a_{11}'\Sigma x_1^2 + a_{12}'\Sigma x_1 x_2 = 1$

$$0.00012198\,(8{,}913.6) + (-0.000056374)(1{,}551.9) = \underline{0.99979}$$

$$b_1 = a_{11}'\Sigma x_1 y + a_{12}'\Sigma x_2 y$$

$$= 0.00012198\,(474.9) + (-0.000056374)(255.3)$$

$$= \underline{0.04354}$$

$$b_2 = a_{12}'\Sigma x_1 y + a_{22}'\Sigma x_2 y$$

$$= (-0.000056374)(474.9) + (0.00032379)(255.3)$$

$$= \underline{0.05589}$$

Multiple Regression Equation—Two Independent Variables:

$$\hat{Y} = \bar{y} + b_1 x_1 + b_2 x_2$$
$$\hat{Y} = 3.67 + 0.04354\,(X_1 - 42.40) + 0.05589\,(X_2 - 23.29)$$
$$\hat{Y} = \underline{0.529 + 0.04354\,X_1 + 0.05589\,X_2}$$

$$R_{y.12} = \sqrt{\frac{\Sigma \hat{y}^2}{\Sigma y^2}} = \sqrt{\frac{34.946}{43.5}}$$

$$R_{y.12} = \underline{0.8963**}$$

The evaluation of the significance of the partial regression coefficients is performed exactly as before with the exception being the use of a_{ij}''s.

$$s_{y.12} = \sqrt{\frac{43.5 - 34.946}{12}} = \underline{0.8443} \text{ inches of sludge depth}$$

$$s_{b_1} = \sqrt{a_{11}' s_{y.12}^2} = \sqrt{0.00012198\,(0.8443)^2}$$

$$s_{b_1} = \underline{0.009325}$$

$$s_{b_2} = \sqrt{a'_{22} s^2_{y.12}} = \sqrt{0.00032379\,(0.8443)^2}$$

$$s_{b_2} = \underline{0.015192}$$

Since the s_{b_i}'s have decreased, it is obvious that the significance level will increase. Therefore, the b_1 and b_2 are both highly significant in estimating the values of Y.

Conclusions:

There is a highly significant relationship between the depth of sludge in the lagoons and the age of the lagoons and the BOD_5 applied per acre per day. The partial regression coefficient for total solids (X_3) when included in the analysis was not significant, indicating this independent variable had little affect on the sludge accumulation in the lagoons.

Appendix

Table A-1

VALUES OF t

df	Probability of a larger value of t, sign ignored								
	0.5	0.4	0.3	0.2	0.1	0.05	0.02	0.01	0.001
1	1.000	1.376	1.963	3.078	6.314	12.706	31.821	63.657	636.619
2	.816	1.061	1.386	1.886	2.920	4.303	6.965	9.925	31.598
3	.765	.978	1.250	1.638	2.353	3.182	4.541	5.841	12.941
4	.741	.941	1.190	1.533	2.132	2.776	3.747	4.604	8.610
5	.727	.920	1.156	1.476	2.015	2.571	3.365	4.032	6.859
6	.718	.906	1.134	1.440	1.943	2.447	3.143	3.707	5.959
7	.711	.896	1.119	1.415	1.895	2.365	2.998	3.499	5.405
8	.706	.889	1.108	1.397	1.860	2.306	2.896	3.355	5.041
9	.703	.883	1.100	1.383	1.833	2.262	2.821	3.250	4.781
10	.700	.879	1.093	1.372	1.812	2.228	2.764	3.169	4.587
11	.697	.876	1.088	1.363	1.796	2.201	2.718	3.106	4.437
12	.695	.873	1.083	1.356	1.782	2.179	2.681	3.055	4.318
13	.694	.870	1.079	1.350	1.771	2.160	2.650	3.012	4.221
14	.692	.868	1.076	1.345	1.761	2.145	2.624	2.977	4.140
15	.691	.866	1.074	1.341	1.753	2.131	2.602	2.947	4.073
16	.690	.865	1.071	1.337	1.746	2.120	2.583	2.921	4.015
17	.689	.863	1.069	1.333	1.740	2.110	2.567	2.898	3.965
18	.688	.862	1.067	1.330	1.734	2.101	2.552	2.878	3.922
19	.688	.861	1.066	1.328	1.729	2.093	2.539	2.861	3.883
20	.687	.860	1.064	1.325	1.725	2.086	2.528	2.845	3.850
21	.686	.859	1.063	1.323	1.721	2.080	2.518	2.831	3.819
22	.686	.858	1.061	1.321	1.717	2.074	2.508	2.819	3.792
23	.685	.858	1.060	1.319	1.714	2.069	2.500	2.807	3.767
24	.685	.857	1.059	1.318	1.711	2.064	2.492	2.797	3.745
25	.684	.856	1.058	1.316	1.708	2.060	2.485	2.787	3.725
26	.684	.856	1.058	1.315	1.706	2.056	2.479	2.779	3.707
27	.684	.855	1.057	1.314	1.703	2.052	2.473	2.771	3.690
28	.683	.855	1.056	1.313	1.701	2.048	2.467	2.763	3.674
29	.683	.854	1.055	1.311	1.699	2.045	2.462	2.756	3.659
30	.683	.854	1.055	1.310	1.697	2.042	2.457	2.750	3.646
40	.681	.851	1.050	1.303	1.684	2.021	2.423	2.704	3.551
60	.679	.848	1.046	1.296	1.671	2.000	2.390	2.660	3.460
120	.677	.845	1.041	1.289	1.658	1.980	2.358	2.617	3.373
∞	.674	.842	1.036	1.282	1.645	1.960	2.326	2.576	3.291
df	0.25	0.2	0.15	0.1	0.05	0.025	0.01	0.005	0.0005
	Probability of a larger value of t, sign considered								

Source: This table is abridged from Table III of Fisher and Yates, "Statistical Tables for Biological, Agricultural, and Medical Research" published by Longman Group Ltd., London, 1974. (Previously published by Oliver and Boyd Ltd., Edinburgh, 1949), by permission of the authors and publishers.

Table A-2

VALUES OF F

Denominator df	Probability of a larger F	Numerator df								
		1	2	3	4	5	6	7	8	9
1	.100	39.86	49.50	53.59	55.83	57.24	58.20	58.91	59.44	59.86
	.050	161.4	199.5	215.7	224.6	230.2	234.0	236.8	238.9	240.5
	.025	647.8	799.5	864.2	899.6	921.8	937.1	948.2	956.7	963.3
	.010	4052	4999.5	5403	5625	5764	5859	5928	5982	6022
	.005	16211	20000	21615	22500	23056	23437	23715	23925	24091
2	.100	8.53	9.00	9.16	9.24	9.29	9.33	9.35	9.37	9.38
	.050	18.51	19.00	19.16	19.25	19.30	19.33	19.35	19.37	19.38
	.025	38.51	39.00	39.17	39.25	39.30	39.33	39.36	39.37	39.39
	.010	98.50	99.00	99.17	99.25	99.30	99.33	99.36	99.37	99.39
	.005	198.5	199.0	199.2	199.2	199.3	199.3	199.4	199.4	199.4
3	.100	5.54	5.46	5.39	5.34	5.31	5.28	5.27	5.25	5.24
	.050	10.13	9.55	9.28	9.12	9.01	8.94	8.89	8.85	8.81
	.025	17.44	16.04	15.44	15.10	14.88	14.73	14.62	14.54	14.47
	.010	34.12	30.82	29.46	28.71	28.24	27.91	27.67	27.49	27.35
	.005	55.55	49.80	47.47	46.19	45.39	44.84	44.43	44.13	43.88
4	.100	4.54	4.32	4.19	4.11	4.05	4.01	3.98	3.95	3.94
	.050	7.71	6.94	6.59	6.39	6.26	6.16	6.09	6.04	6.00
	.025	12.22	10.65	9.98	9.60	9.36	9.20	9.07	8.98	8.90
	.010	21.20	18.00	16.69	15.98	15.52	15.21	14.98	14.80	14.66
	.005	31.33	26.28	24.26	23.15	22.46	21.97	21.62	21.35	21.14
5	.100	4.06	3.78	3.62	3.52	3.45	3.40	3.37	3.34	3.32
	.050	6.61	5.79	5.41	5.19	5.05	4.95	4.88	4.82	4.77
	.025	10.01	8.43	7.76	7.39	7.15	6.98	6.85	6.76	6.68
	.010	16.26	13.27	12.06	11.39	10.97	10.67	10.46	10.29	10.16
	.005	22.78	18.31	16.53	15.56	14.94	14.51	14.20	13.96	13.77
6	.100	3.78	3.46	3.29	3.18	3.11	3.05	3.01	2.98	2.96
	.050	5.99	5.14	4.76	4.53	4.39	4.28	4.21	4.15	4.10
	.025	8.81	7.26	6.60	6.23	5.99	5.82	5.70	5.60	5.52
	.010	13.75	10.92	9.78	9.15	8.75	8.47	8.26	8.10	7.98
	.005	18.63	14.54	12.92	12.03	11.46	11.07	10.79	10.57	10.39
7	.100	3.59	3.26	3.07	2.96	2.88	2.83	2.78	2.75	2.72
	.050	5.59	4.74	4.35	4.12	3.97	3.87	3.79	3.73	3.68
	.025	8.07	6.54	5.89	5.52	5.29	5.12	4.99	4.90	4.82
	.010	12.25	9.55	8.45	7.85	7.46	7.19	6.99	6.84	6.72
	.005	16.24	12.40	10.88	10.05	9.52	9.16	8.89	8.68	8.51
8	.100	3.46	3.11	2.92	2.81	2.73	2.67	2.62	2.59	2.56
	.050	5.32	4.46	4.07	3.84	3.69	3.58	3.50	3.44	3.39
	.025	7.57	6.06	5.42	5.05	4.82	4.65	4.53	4.43	4.36
	.010	11.26	8.65	7.59	7.01	6.63	6.37	6.18	6.03	5.91
	.005	14.69	11.04	9.60	8.81	8.30	7.95	7.69	7.50	7.34
9	.100	3.36	3.01	2.81	2.69	2.61	2.55	2.51	2.47	2.44
	.050	5.12	4.26	3.86	3.63	3.48	3.37	3.29	3.23	3.18
	.025	7.21	5.71	5.08	4.72	4.48	4.32	4.20	4.10	4.03
	.010	10.56	8.02	6.99	6.42	6.06	5.80	5.61	5.47	5.35
	.005	13.61	10.11	8.72	7.96	7.47	7.13	6.88	6.69	6.54
10	.100	3.29	2.92	2.73	2.61	2.52	2.46	2.41	2.38	2.35
	.050	4.96	4.10	3.71	3.48	3.33	3.22	3.14	3.07	3.02
	.025	6.94	5.46	4.83	4.47	4.24	4.07	3.95	3.85	3.78
	.010	10.04	7.56	6.55	5.99	5.64	5.39	5.20	5.06	4.94
	.005	12.83	9.43	8.08	7.34	6.87	6.54	6.30	6.12	5.97
11	.100	3.23	2.86	2.66	2.54	2.45	2.39	2.34	2.30	2.27
	.050	4.84	3.98	3.59	3.36	3.20	3.09	3.01	2.95	2.90
	.025	6.72	5.26	4.63	4.28	4.04	3.88	3.76	3.66	3.59
	.010	9.65	7.21	6.22	5.67	5.32	5.07	4.89	4.74	4.63
	.005	12.23	8.91	7.60	6.88	6.42	6.10	5.86	5.68	5.54
12	.100	3.18	2.81	2.61	2.48	2.39	2.33	2.28	2.24	2.21
	.050	4.75	3.89	3.49	3.26	3.11	3.00	2.91	2.85	2.80
	.025	6.55	5.10	4.47	4.12	3.89	3.73	3.61	3.51	3.44
	.010	9.33	6.93	5.95	5.41	5.06	4.82	4.64	4.50	4.39
	.005	11.75	8.51	7.23	6.52	6.07	5.76	5.52	5.35	5.20
13	.100	3.14	2.76	2.56	2.43	2.35	2.28	2.23	2.20	2.16
	.050	4.67	3.81	3.41	3.18	3.03	2.92	2.83	2.77	2.71
	.025	6.41	4.97	4.35	4.00	3.77	3.60	3.48	3.39	3.31
	.010	9.07	6.70	5.74	5.21	4.86	4.62	4.44	4.30	4.19
	.005	11.37	8.19	6.93	6.23	5.79	5.48	5.25	5.08	4.94
14	.100	3.10	2.73	2.52	2.39	2.31	2.24	2.19	2.15	2.12
	.050	4.60	3.74	3.34	3.11	2.96	2.85	2.76	2.70	2.65
	.025	6.30	4.86	4.24	3.89	3.66	3.50	3.38	3.29	3.21
	.010	8.86	6.51	5.56	5.04	4.69	4.46	4.28	4.14	4.03
	.005	11.06	7.92	6.68	6.00	5.56	5.26	5.03	4.86	4.72

Table A-2. Continued

VALUES OF F

			Numerator df									
10	12	15	20	24	30	40	60	120	∞	P	df
60.19	60.71	61.22	61.74	62.00	62.26	62.53	62.79	63.06	63.33	.100	1
241.9	243.9	245.9	248.0	249.1	250.1	251.1	252.2	253.3	254.3	.050	
968.6	976.7	984.9	993.1	997.2	1001	1006	1010	1014	1018	.025	
6056	6106	6157	6209	6235	6261	6287	6313	6339	6366	.010	
24224	24426	24630	24836	24940	25044	25148	25253	25359	25465	.005	
9.39	9.41	9.42	9.44	9.45	9.46	9.47	9.47	9.48	9.49	.100	2
19.40	19.41	19.43	19.45	19.45	19.46	19.47	19.48	19.49	19.50	.050	
39.40	39.41	39.43	39.45	39.46	39.46	39.47	39.48	39.49	39.50	.025	
99.40	99.42	99.43	99.45	99.46	99.47	99.47	99.48	99.49	99.50	.010	
199.4	199.4	199.4	199.4	199.5	199.5	199.5	199.5	199.5	199.5	.005	
5.23	5.22	5.20	5.18	5.18	5.17	5.16	5.15	5.14	5.13	.100	3
8.79	8.74	8.70	8.66	8.64	8.62	8.59	8.57	8.55	8.53	.050	
14.42	14.34	14.25	14.17	14.12	14.08	14.04	13.99	13.95	13.90	.025	
27.23	27.05	26.87	26.69	26.60	26.50	26.41	26.32	26.22	26.13	.010	
43.69	43.39	43.08	42.78	42.62	42.47	42.31	42.15	41.99	41.83	.005	
3.92	3.90	3.87	3.84	3.83	3.82	3.80	3.79	3.78	3.76	.100	4
5.96	5.91	5.86	5.80	5.77	5.75	5.72	5.69	5.66	5.63	.050	
8.84	8.75	8.66	8.56	8.51	8.46	8.41	8.36	8.31	8.26	.025	
14.55	14.37	14.20	14.02	13.93	13.84	13.75	13.65	13.56	13.46	.010	
20.97	20.70	20.44	20.17	20.03	19.89	19.75	19.61	19.47	19.32	.005	
3.30	3.27	3.24	3.21	3.19	3.17	3.16	3.14	3.12	3.10	.100	5
4.74	4.68	4.62	4.56	4.53	4.50	4.46	4.43	4.40	4.36	.050	
6.62	6.52	6.43	6.33	6.28	6.23	6.18	6.12	6.07	6.02	.025	
10.05	9.89	9.72	9.55	9.47	9.38	9.29	9.20	9.11	9.02	.010	
13.62	13.38	13.15	12.90	12.78	12.66	12.53	12.40	12.27	12.14	.005	
2.94	2.90	2.87	2.84	2.82	2.80	2.78	2.76	2.74	2.72	.100	6
4.06	4.00	3.94	3.87	3.84	3.81	3.77	3.74	3.70	3.67	.050	
5.46	5.37	5.27	5.17	5.12	5.07	5.01	4.96	4.90	4.85	.025	
7.87	7.72	7.56	7.40	7.31	7.23	7.14	7.06	6.97	6.88	.010	
10.25	10.03	9.81	9.59	9.47	9.36	9.24	9.12	9.00	8.88	.005	
2.70	2.67	2.63	2.59	2.58	2.56	2.54	2.51	2.49	2.47	.100	7
3.64	3.57	3.51	3.44	3.41	3.38	3.34	3.30	3.27	3.23	.050	
4.76	4.67	4.57	4.47	4.42	4.36	4.31	4.25	4.20	4.14	.025	
6.62	6.47	6.31	6.16	6.07	5.99	5.91	5.82	5.74	5.65	.010	
8.38	8.18	7.97	7.75	7.65	7.53	7.42	7.31	7.19	7.08	.005	
2.54	2.50	2.46	2.42	2.40	2.38	2.36	2.34	2.32	2.29	.100	8
3.35	3.28	3.22	3.15	3.12	3.08	3.04	3.01	2.97	2.93	.050	
4.30	4.20	4.10	4.00	3.95	3.89	3.84	3.78	3.73	3.67	.025	
5.81	5.67	5.52	5.36	5.28	5.20	5.12	5.03	4.95	4.86	.010	
7.21	7.01	6.81	6.61	6.50	6.40	6.29	6.18	6.06	5.95	.005	
2.42	2.38	2.34	2.30	2.28	2.25	2.23	2.21	2.18	2.16	.100	9
3.14	3.07	3.01	2.94	2.90	2.86	2.83	2.79	2.75	2.71	.050	
3.96	3.87	3.77	3.67	3.61	3.56	3.51	3.45	3.39	3.33	.025	
5.26	5.11	4.96	4.81	4.73	4.65	4.57	4.48	4.40	4.31	.010	
6.42	6.23	6.03	5.83	5.73	5.62	5.52	5.41	5.30	5.19	.005	
2.32	2.28	2.24	2.20	2.18	2.16	2.13	2.11	2.08	2.06	.100	10
2.98	2.91	2.85	2.77	2.74	2.70	2.66	2.62	2.58	2.54	.050	
3.72	3.62	3.52	3.42	3.37	3.31	3.26	3.20	3.14	3.08	.025	
4.85	4.71	4.56	4.41	4.33	4.25	4.17	4.08	4.00	3.91	.010	
5.85	5.66	5.47	5.27	5.17	5.07	4.97	4.86	4.75	4.64	.005	
2.25	2.21	2.17	2.12	2.10	2.08	2.05	2.03	2.00	1.97	.100	11
2.85	2.79	2.72	2.65	2.61	2.57	2.53	2.49	2.45	2.40	.050	
3.53	3.43	3.33	3.23	3.17	3.12	3.06	3.00	2.94	2.88	.025	
4.54	4.40	4.25	4.10	4.02	3.94	3.86	3.78	3.69	3.60	.010	
5.42	5.24	5.05	4.86	4.76	4.65	4.55	4.44	4.34	4.23	.005	
2.19	2.15	2.10	2.06	2.04	2.01	1.99	1.96	1.93	1.90	.100	12
2.75	2.69	2.62	2.54	2.51	2.47	2.43	2.38	2.34	2.30	.050	
3.37	3.28	3.18	3.07	3.02	2.96	2.91	2.85	2.79	2.72	.025	
4.30	4.16	4.01	3.86	3.78	3.70	3.62	3.54	3.45	3.36	.010	
5.09	4.91	4.72	4.53	4.43	4.33	4.23	4.12	4.01	3.90	.005	
2.14	2.10	2.05	2.01	1.98	1.96	1.93	1.90	1.88	1.85	.100	13
2.67	2.60	2.53	2.46	2.42	2.38	2.34	2.30	2.25	2.21	.050	
3.25	3.15	3.05	2.95	2.89	2.84	2.78	2.72	2.66	2.60	.025	
4.10	3.96	3.82	3.66	3.59	3.51	3.43	3.34	3.25	3.17	.010	
4.82	4.64	4.46	4.27	4.17	4.07	3.97	3.87	3.76	3.65	.005	
2.10	2.05	2.01	1.96	1.94	1.91	1.89	1.86	1.83	1.80	.100	14
2.60	2.53	2.46	2.39	2.35	2.31	2.27	2.22	2.18	2.13	.050	
3.15	3.05	2.95	2.84	2.79	2.73	2.67	2.61	2.55	2.49	.025	
3.94	3.80	3.66	3.51	3.43	3.35	3.27	3.18	3.09	3.00	.010	
4.60	4.43	4.25	4.06	3.96	3.86	3.76	3.66	3.55	3.44	.005	

Table A-2. Continued

VALUES OF F

Denominator df	Probability of a larger F	Numerator df								
		1	2	3	4	5	6	7	8	9
15	.100	3.07	2.70	2.49	2.36	2.27	2.21	2.16	2.12	2.09
	.050	4.54	3.68	3.29	3.06	2.90	2.79	2.71	2.64	2.59
	.025	6.20	4.77	4.15	3.80	3.58	3.41	3.29	3.20	3.12
	.010	8.68	6.36	5.42	4.89	4.56	4.32	4.14	4.00	3.89
	.005	10.80	7.70	6.48	5.80	5.37	5.07	4.85	4.67	4.54
16	.100	3.05	2.67	2.46	2.33	2.24	2.18	2.13	2.09	2.06
	.050	4.49	3.63	3.24	3.01	2.85	2.74	2.66	2.59	2.54
	.025	6.12	4.69	4.08	3.73	3.50	3.34	3.22	3.12	3.05
	.010	8.53	6.23	5.29	4.77	4.44	4.20	4.03	3.89	3.78
	.005	10.58	7.51	6.30	5.64	5.21	4.91	4.69	4.52	4.38
17	.100	3.03	2.64	2.44	2.31	2.22	2.15	2.10	2.06	2.03
	.050	4.45	3.59	3.20	2.96	2.81	2.70	2.61	2.55	2.49
	.025	6.04	4.62	4.01	3.66	3.44	3.28	3.16	3.06	2.98
	.010	8.40	6.11	5.18	4.67	4.34	4.10	3.93	3.79	3.68
	.005	10.38	7.35	6.16	5.50	5.07	4.78	4.56	4.39	4.25
18	.100	3.01	2.62	2.42	2.29	2.20	2.13	2.08	2.04	2.00
	.050	4.41	3.55	3.16	2.93	2.77	2.66	2.58	2.51	2.46
	.025	5.98	4.56	3.95	3.61	3.38	3.22	3.10	3.01	2.93
	.010	8.29	6.01	5.09	4.58	4.25	4.01	3.84	3.71	3.60
	.005	10.22	7.21	6.03	5.37	4.96	4.66	4.44	4.28	4.14
19	.100	2.99	2.61	2.40	2.27	2.18	2.11	2.06	2.02	1.98
	.050	4.38	3.52	3.13	2.90	2.74	2.63	2.54	2.48	2.42
	.025	5.92	4.51	3.90	3.56	3.33	3.17	3.05	2.96	2.88
	.010	8.18	5.93	5.01	4.50	4.17	3.94	3.77	3.63	3.52
	.005	10.07	7.09	5.92	5.27	4.85	4.56	4.34	4.18	4.04
20	.100	2.97	2.59	2.38	2.25	2.16	2.09	2.04	2.00	1.96
	.050	4.35	3.49	3.10	2.87	2.71	2.60	2.51	2.45	2.39
	.025	5.87	4.46	3.86	3.51	3.29	3.13	3.01	2.91	2.84
	.010	8.10	5.85	4.94	4.43	4.10	3.87	3.70	3.56	3.46
	.005	9.94	6.99	5.82	5.17	4.76	4.47	4.26	4.09	3.96
21	.100	2.96	2.57	2.36	2.23	2.14	2.08	2.02	1.98	1.95
	.050	4.32	3.47	3.07	2.84	2.68	2.57	2.49	2.42	2.37
	.025	5.83	4.42	3.82	3.48	3.25	3.09	2.97	2.87	2.80
	.010	8.02	5.78	4.87	4.37	4.04	3.81	3.64	3.51	3.40
	.005	9.83	6.89	5.73	5.09	4.68	4.39	4.18	4.01	3.88
22	.100	2.95	2.56	2.35	2.22	2.13	2.06	2.01	1.97	1.93
	.050	4.30	3.44	3.05	2.82	2.66	2.55	2.46	2.40	2.34
	.025	5.79	4.38	3.78	3.44	3.22	3.05	2.93	2.84	2.76
	.010	7.95	5.72	4.82	4.31	3.99	3.76	3.59	3.45	3.35
	.005	9.73	6.81	5.65	5.02	4.61	4.32	4.11	3.94	3.81
23	.100	2.94	2.55	2.34	2.21	2.11	2.05	1.99	1.95	1.92
	.050	4.28	3.42	3.03	2.80	2.64	2.53	2.44	2.37	2.32
	.025	5.75	4.35	3.75	3.41	3.18	3.02	2.90	2.81	2.73
	.010	7.88	5.66	4.76	4.26	3.94	3.71	3.54	3.41	3.30
	.005	9.63	6.73	5.58	4.95	4.54	4.26	4.05	3.88	3.75
24	.100	2.93	2.54	2.33	2.19	2.10	2.04	1.98	1.94	1.91
	.050	4.26	3.40	3.01	2.78	2.62	2.51	2.42	2.36	2.30
	.025	5.72	4.32	3.72	3.38	3.15	2.99	2.87	2.78	2.70
	.010	7.82	5.61	4.72	4.22	3.90	3.67	3.50	3.36	3.26
	.005	9.55	6.66	5.52	4.89	4.49	4.20	3.99	3.83	3.69
25	.100	2.92	2.53	2.32	2.18	2.09	2.02	1.97	1.93	1.89
	.050	4.24	3.39	2.99	2.76	2.60	2.49	2.40	2.34	2.28
	.025	5.69	4.29	3.69	3.35	3.13	2.97	2.85	2.75	2.68
	.010	7.77	5.57	4.68	4.18	3.85	3.63	3.46	3.32	3.22
	.005	9.48	6.60	5.46	4.84	4.43	4.15	3.94	3.78	3.64
26	.100	2.91	2.52	2.31	2.17	2.08	2.01	1.96	1.92	1.88
	.050	4.23	3.37	2.98	2.74	2.59	2.47	2.39	2.32	2.27
	.025	5.66	4.27	3.67	3.33	3.10	2.94	2.82	2.73	2.65
	.010	7.72	5.53	4.64	4.14	3.82	3.59	3.42	3.29	3.18
	.005	9.41	6.54	5.41	4.79	4.38	4.10	3.89	3.73	3.60
27	.100	2.90	2.51	2.30	2.17	2.07	2.00	1.95	1.91	1.87
	.050	4.21	3.35	2.96	2.73	2.57	2.46	2.37	2.31	2.25
	.025	5.63	4.24	3.65	3.31	3.08	2.92	2.80	2.71	2.63
	.010	7.68	5.49	4.60	4.11	3.78	3.56	3.39	3.26	3.15
	.005	9.34	6.49	5.36	4.74	4.34	4.06	3.85	3.69	3.56
28	.100	2.89	2.50	2.29	2.16	2.06	2.00	1.94	1.90	1.87
	.050	4.20	3.34	2.95	2.71	2.56	2.45	2.36	2.29	2.24
	.025	5.61	4.22	3.63	3.29	3.06	2.90	2.78	2.69	2.61
	.010	7.64	5.45	4.57	4.07	3.75	3.53	3.36	3.23	3.12
	.005	9.28	6.44	5.32	4.70	4.30	4.02	3.81	3.65	3.52

Table A-2. Continued

VALUES OF F

10	12	15	20	24	30	40	60	120	∞	P	df
2.06	2.02	1.97	1.92	1.90	1.87	1.85	1.82	1.79	1.76	.100	15
2.54	2.48	2.40	2.33	2.29	2.25	2.20	2.16	2.11	2.07	.050	
3.06	2.96	2.86	2.76	2.70	2.64	2.59	2.52	2.46	2.40	.025	
3.80	3.67	3.52	3.37	3.29	3.21	3.13	3.05	2.96	2.87	.010	
4.42	4.25	4.07	3.88	3.79	3.69	3.58	3.48	3.37	3.26	.005	
2.03	1.99	1.94	1.89	1.87	1.84	1.81	1.78	1.75	1.72	.100	16
2.49	2.42	2.35	2.28	2.24	2.19	2.15	2.11	2.06	2.01	.050	
2.99	2.89	2.79	2.68	2.63	2.57	2.51	2.45	2.38	2.32	.025	
3.69	3.55	3.41	3.26	3.18	3.10	3.02	2.93	2.84	2.75	.010	
4.27	4.10	3.92	3.73	3.64	3.54	3.44	3.33	3.22	3.11	.005	
2.00	1.96	1.91	1.86	1.84	1.81	1.78	1.75	1.72	1.69	.100	17
2.45	2.38	2.31	2.23	2.19	2.15	2.10	2.06	2.01	1.96	.050	
2.92	2.82	2.72	2.62	2.56	2.50	2.44	2.38	2.32	2.25	.025	
3.59	3.46	3.31	3.16	3.08	3.00	2.92	2.83	2.75	2.65	.010	
4.14	3.97	3.79	3.61	3.51	3.41	3.31	3.21	3.10	2.98	.005	
1.98	1.93	1.89	1.84	1.81	1.78	1.75	1.72	1.69	1.66	.100	18
2.41	2.34	2.27	2.19	2.15	2.11	2.06	2.02	1.97	1.92	.050	
2.87	2.77	2.67	2.56	2.50	2.44	2.38	2.32	2.26	2.19	.025	
3.51	3.37	3.23	3.08	3.00	2.92	2.84	2.75	2.66	2.57	.010	
4.03	3.86	3.68	3.50	3.40	3.30	3.20	3.10	2.99	2.87	.005	
1.96	1.91	1.86	1.81	1.79	1.76	1.73	1.70	1.67	1.63	.100	19
2.38	2.31	2.23	2.16	2.11	2.07	2.03	1.98	1.93	1.88	.050	
2.82	2.72	2.62	2.51	2.45	2.39	2.33	2.27	2.20	2.13	.025	
3.43	3.30	3.15	3.00	2.92	2.84	2.76	2.67	2.58	2.49	.010	
3.93	3.76	3.59	3.40	3.31	3.21	3.11	3.00	2.89	2.78	.005	
1.94	1.89	1.84	1.79	1.77	1.74	1.71	1.68	1.64	1.61	.100	20
2.35	2.28	2.20	2.12	2.08	2.04	1.99	1.95	1.90	1.84	.050	
2.77	2.68	2.57	2.46	2.41	2.35	2.29	2.22	2.16	2.09	.025	
3.37	3.23	3.09	2.94	2.86	2.78	2.69	2.61	2.52	2.42	.010	
3.85	3.68	3.50	3.32	3.22	3.12	3.02	2.92	2.81	2.69	.005	
1.92	1.87	1.83	1.78	1.75	1.72	1.69	1.66	1.62	1.59	.100	21
2.32	2.25	2.18	2.10	2.05	2.01	1.96	1.92	1.87	1.81	.050	
2.73	2.64	2.53	2.42	2.37	2.31	2.25	2.18	2.11	2.04	.025	
3.31	3.17	3.03	2.88	2.80	2.72	2.64	2.55	2.46	2.36	.010	
3.77	3.60	3.43	3.24	3.15	3.05	2.95	2.84	2.73	2.61	.005	
1.90	1.86	1.81	1.76	1.73	1.70	1.67	1.64	1.60	1.57	.100	22
2.30	2.23	2.15	2.07	2.03	1.98	1.94	1.89	1.84	1.78	.050	
2.70	2.60	2.50	2.39	2.33	2.27	2.21	2.14	2.08	2.00	.025	
3.26	3.12	2.98	2.83	2.75	2.67	2.58	2.50	2.40	2.31	.010	
3.70	3.54	3.36	3.18	3.08	2.98	2.88	2.77	2.66	2.55	.005	
1.89	1.84	1.80	1.74	1.72	1.69	1.66	1.62	1.59	1.55	.100	23
2.27	2.20	2.13	2.05	2.01	1.96	1.91	1.86	1.81	1.76	.050	
2.67	2.57	2.47	2.36	2.30	2.24	2.18	2.11	2.04	1.97	.025	
3.21	3.07	2.93	2.78	2.70	2.62	2.54	2.45	2.35	2.26	.010	
3.64	3.47	3.30	3.12	3.02	2.92	2.82	2.71	2.60	2.48	.005	
1.88	1.83	1.78	1.73	1.70	1.67	1.64	1.61	1.57	1.53	.100	24
2.25	2.18	2.11	2.03	1.98	1.94	1.89	1.84	1.79	1.73	.050	
2.64	2.54	2.44	2.33	2.27	2.21	2.15	2.08	2.01	1.94	.025	
3.17	3.03	2.89	2.74	2.66	2.58	2.49	2.40	2.31	2.21	.010	
3.59	3.42	3.25	3.06	2.97	2.87	2.77	2.66	2.55	2.43	.005	
1.87	1.82	1.77	1.72	1.69	1.66	1.63	1.59	1.56	1.52	.100	25
2.24	2.16	2.09	2.01	1.96	1.92	1.87	1.82	1.77	1.71	.050	
2.61	2.51	2.41	2.30	2.24	2.18	2.12	2.05	1.98	1.91	.025	
3.13	2.99	2.85	2.70	2.62	2.54	2.45	2.36	2.27	2.17	.010	
3.54	3.37	3.20	3.01	2.92	2.82	2.72	2.61	2.50	2.38	.005	
1.86	1.81	1.76	1.71	1.68	1.65	1.61	1.58	1.54	1.50	.100	26
2.22	2.15	2.07	1.99	1.95	1.90	1.85	1.80	1.75	1.69	.050	
2.59	2.49	2.39	2.28	2.22	2.16	2.09	2.03	1.95	1.88	.025	
3.09	2.96	2.81	2.66	2.58	2.50	2.42	2.33	2.23	2.13	.010	
3.49	3.33	3.15	2.97	2.87	2.77	2.67	2.56	2.45	2.33	.005	
1.85	1.80	1.75	1.70	1.67	1.64	1.60	1.57	1.53	1.49	.100	27
2.20	2.13	2.06	1.97	1.93	1.88	1.84	1.79	1.73	1.67	.050	
2.57	2.47	2.36	2.25	2.19	2.13	2.07	2.00	1.93	1.85	.025	
3.06	2.93	2.78	2.63	2.55	2.47	2.38	2.29	2.20	2.10	.010	
3.45	3.28	3.11	2.93	2.83	2.73	2.63	2.52	2.41	2.29	.005	
1.84	1.79	1.74	1.69	1.66	1.63	1.59	1.56	1.52	1.48	.100	28
2.19	2.12	2.04	1.96	1.91	1.87	1.82	1.77	1.71	1.65	.050	
2.55	2.45	2.34	2.23	2.17	2.11	2.05	1.98	1.91	1.83	.025	
3.03	2.90	2.75	2.60	2.52	2.44	2.35	2.26	2.17	2.06	.010	
3.41	3.25	3.07	2.89	2.79	2.69	2.59	2.48	2.37	2.25	.005	

Table A-2. Continued

VALUES OF F

Denominator df	Probability of a larger F	1	2	3	4	5	6	7	8	9
29	.100	2.89	2.50	2.28	2.15	2.06	1.99	1.93	1.89	1.86
	.050	4.18	3.33	2.93	2.70	2.55	2.43	2.35	2.28	2.22
	.025	5.59	4.20	3.61	3.27	3.04	2.88	2.76	2.67	2.59
	.010	7.60	5.42	4.54	4.04	3.73	3.50	3.33	3.20	3.09
	.005	9.23	6.40	5.28	4.66	4.26	3.98	3.77	3.61	3.48
30	.100	2.88	2.49	2.28	2.14	2.05	1.98	1.93	1.88	1.85
	.050	4.17	3.32	2.92	2.69	2.53	2.42	2.33	2.27	2.21
	.025	5.57	4.18	3.59	3.25	3.03	2.87	2.75	2.65	2.57
	.010	7.56	5.39	4.51	4.02	3.70	3.47	3.30	3.17	3.07
	.005	9.18	6.35	5.24	4.62	4.23	3.95	3.74	3.58	3.45
40	.100	2.84	2.44	2.23	2.09	2.00	1.93	1.87	1.83	1.79
	.050	4.08	3.23	2.84	2.61	2.45	2.34	2.25	2.18	2.12
	.025	5.42	4.05	3.46	3.13	2.90	2.74	2.62	2.53	2.45
	.010	7.31	5.18	4.31	3.83	3.51	3.29	3.12	2.99	2.89
	.005	8.83	6.07	4.98	4.37	3.99	3.71	3.51	3.35	3.22
60	.100	2.79	2.39	2.18	2.04	1.95	1.87	1.82	1.77	1.74
	.050	4.00	3.15	2.76	2.53	2.37	2.25	2.17	2.10	2.04
	.025	5.29	3.93	3.34	3.01	2.79	2.63	2.51	2.41	2.33
	.010	7.08	4.98	4.13	3.65	3.34	3.12	2.95	2.82	2.72
	.005	8.49	5.79	4.73	4.14	3.76	3.49	3.29	3.13	3.01
120	.100	2.75	2.35	2.13	1.99	1.90	1.82	1.77	1.72	1.68
	.050	3.92	3.07	2.68	2.45	2.29	2.17	2.09	2.02	1.96
	.025	5.15	3.80	3.23	2.89	2.67	2.52	2.39	2.30	2.22
	.010	6.85	4.79	3.95	3.48	3.17	2.96	2.79	2.66	2.56
	.005	8.18	5.54	4.50	3.92	3.55	3.28	3.09	2.93	2.81
∞	.100	2.71	2.30	2.08	1.94	1.85	1.77	1.72	1.67	1.63
	.050	3.84	3.00	2.60	2.37	2.21	2.10	2.01	1.94	1.88
	.025	5.02	3.69	3.12	2.79	2.57	2.41	2.29	2.19	2.11
	.010	6.63	4.61	3.78	3.32	3.02	2.80	2.64	2.51	2.41
	.005	7.88	5.30	4.28	3.72	3.35	3.09	2.90	2.74	2.62

SOURCE: A portion of "Tables of percentage points of the inverted beta (F) distribution," *Biometrika*, vol. 33 (1943) by M. Merrington and C. M. Thompson and from Table 18 of *Biometrika Tables for Statisticians*, vol. 1, Cambridge University Press, 1954, edited by E. S. Pearson and H. O. Hartley. Reproduced with permission of the authors, editors, and *Biometrika* trustees.

Table A-2. Continued

VALUES OF F

Numerator df												
10	12	15	20	24	30	40	60	120	∞	P	df
1.83	1.78	1.73	1.68	1.65	1.62	1.58	1.55	1.51	1.47	.100	29
2.18	2.10	2.03	1.94	1.90	1.85	1.81	1.75	1.70	1.64	.050	
2.53	2.43	2.32	2.21	2.15	2.09	2.03	1.96	1.89	1.81	.025	
3.00	2.87	2.73	2.57	2.49	2.41	2.33	2.23	2.14	2.03	.010	
3.38	3.21	3.04	2.86	2.76	2.66	2.56	2.45	2.33	2.21	.005	
1.82	1.77	1.72	1.67	1.64	1.61	1.57	1.54	1.50	1.46	.100	30
2.16	2.09	2.01	1.93	1.89	1.84	1.79	1.74	1.68	1.62	.050	
2.51	2.41	2.31	2.20	2.14	2.07	2.01	1.94	1.87	1.79	.025	
2.98	2.84	2.70	2.55	2.47	2.39	2.30	2.21	2.11	2.01	.010	
3.34	3.18	3.01	2.82	2.73	2.63	2.52	2.42	2.30	2.18	.005	
1.76	1.71	1.66	1.61	1.57	1.54	1.51	1.47	1.42	1.38	.100	40
2.08	2.00	1.92	1.84	1.79	1.74	1.69	1.64	1.58	1.51	.050	
2.39	2.29	2.18	2.07	2.01	1.94	1.88	1.80	1.72	1.64	.025	
2.80	2.66	2.52	2.37	2.29	2.20	2.11	2.02	1.92	1.80	.010	
3.12	2.95	2.78	2.60	2.50	2.40	2.30	2.18	2.06	1.93	.005	
1.71	1.66	1.60	1.54	1.51	1.48	1.44	1.40	1.35	1.29	.100	60
1.99	1.92	1.84	1.75	1.70	1.65	1.59	1.53	1.47	1.39	.050	
2.27	2.17	2.06	1.94	1.88	1.82	1.74	1.67	1.58	1.48	.025	
2.63	2.50	2.35	2.20	2.12	2.03	1.94	1.84	1.73	1.60	.010	
2.90	2.74	2.57	2.39	2.29	2.19	2.08	1.96	1.83	1.69	.005	
1.65	1.60	1.55	1.48	1.45	1.41	1.37	1.32	1.26	1.19	.100	120
1.91	1.83	1.75	1.66	1.61	1.55	1.50	1.43	1.35	1.25	.050	
2.16	2.05	1.94	1.82	1.76	1.69	1.61	1.53	1.43	1.31	.025	
2.47	2.34	2.19	2.03	1.95	1.86	1.76	1.66	1.53	1.38	.010	
2.71	2.54	2.37	2.19	2.09	1.98	1.87	1.75	1.61	1.43	.005	
1.60	1.55	1.49	1.42	1.38	1.34	1.30	1.24	1.17	1.00	.100	∞
1.83	1.75	1.67	1.57	1.52	1.46	1.39	1.32	1.22	1.00	.050	
2.05	1.94	1.83	1.71	1.64	1.57	1.48	1.39	1.27	1.00	.025	
2.32	2.18	2.04	1.88	1.79	1.70	1.59	1.47	1.32	1.00	.010	
2.52	2.36	2.19	2.00	1.90	1.79	1.67	1.53	1.36	1.00	.005	

Table A-3

SIGNIFICANT STUDENTIZED RANGES FOR 5% AND 1% LEVEL NEW MULTIPLE-RANGE TEST

Error df	Protection level	p = number of means for range being tested													
		2	3	4	5	6	7	8	9	10	12	14	16	18	20
16	.05	3.00	3.15	3.23	3.30	3.34	3.37	3.39	3.41	3.43	3.44	3.45	3.46	3.47	3.47
	.01	4.13	4.34	4.45	4.54	4.60	4.67	4.72	4.76	4.79	4.84	4.88	4.91	4.93	4.94
17	.05	2.98	3.13	3.22	3.28	3.33	3.36	3.38	3.40	3.42	3.44	3.45	3.46	3.47	3.47
	.01	4.10	4.30	4.41	4.50	4.56	4.63	4.68	4.72	4.75	4.80	4.83	4.86	4.88	4.89
18	.05	2.97	3.12	3.21	3.27	3.32	3.35	3.37	3.39	3.41	3.43	3.45	3.46	3.47	3.47
	.01	4.07	4.27	4.38	4.46	4.53	4.59	4.64	4.68	4.71	4.76	4.79	4.82	4.84	4.85
19	.05	2.96	3.11	3.19	3.26	3.31	3.35	3.37	3.39	3.41	3.43	3.44	3.46	3.47	3.47
	.01	4.05	4.24	4.35	4.43	4.50	4.56	4.61	4.64	4.67	4.72	4.76	4.79	4.81	4.82
20	.05	2.95	3.10	3.18	3.25	3.30	3.34	3.36	3.38	3.40	3.43	3.44	3.46	3.46	3.47
	.01	4.02	4.22	4.33	4.40	4.47	4.53	4.58	4.61	4.65	4.69	4.73	4.76	4.78	4.79
22	.05	2.93	3.08	3.17	3.24	3.29	3.32	3.35	3.37	3.39	3.42	3.44	3.45	3.46	3.47
	.01	3.99	4.17	4.28	4.36	4.42	4.48	4.53	4.57	4.60	4.65	4.68	4.71	4.74	4.75
24	.05	2.92	3.07	3.15	3.22	3.28	3.31	3.34	3.37	3.38	3.41	3.44	3.45	3.46	3.47
	.01	3.96	4.14	4.24	4.33	4.39	4.44	4.49	4.53	4.57	4.62	4.64	4.67	4.70	4.72
26	.05	2.91	3.06	3.14	3.21	3.27	3.30	3.34	3.36	3.38	3.41	3.43	3.45	3.46	3.47
	.01	3.93	4.11	4.21	4.30	4.36	4.41	4.46	4.50	4.53	4.58	4.62	4.65	4.67	4.69
28	.05	2.90	3.04	3.13	3.20	3.26	3.30	3.33	3.35	3.37	3.40	3.43	3.45	3.46	3.47
	.01	3.91	4.08	4.18	4.28	4.34	4.39	4.43	4.47	4.51	4.56	4.60	4.62	4.65	4.67
30	.05	2.89	3.04	3.12	3.20	3.25	3.29	3.32	3.35	3.37	3.40	3.43	3.44	3.46	3.47
	.01	3.89	4.06	4.16	4.22	4.32	4.36	4.41	4.45	4.48	4.54	4.58	4.61	4.63	4.65
40	.05	2.86	3.01	3.10	3.17	3.22	3.27	3.30	3.33	3.35	3.39	3.42	3.44	3.46	3.47
	.01	3.82	3.99	4.10	4.17	4.24	4.30	4.34	4.37	4.41	4.46	4.51	4.54	4.57	4.59
60	.05	2.83	2.98	3.08	3.14	3.20	3.24	3.28	3.31	3.33	3.37	3.40	3.43	3.45	3.47
	.01	3.76	3.92	4.03	4.12	4.17	4.23	4.27	4.31	4.34	4.39	4.44	4.47	4.50	4.53
100	.05	2.80	2.95	3.05	3.12	3.18	3.22	3.26	3.29	3.32	3.36	3.40	3.42	3.45	3.47
	.01	3.71	3.86	3.98	4.06	4.11	4.17	4.21	4.25	4.29	4.35	4.38	4.42	4.45	4.48
∞	.05	2.77	2.92	3.02	3.09	3.15	3.19	3.23	3.26	3.29	3.34	3.38	3.41	3.44	3.47
	.01	3.64	3.80	3.90	3.98	4.04	4.09	4.14	4.17	4.20	4.26	4.31	4.34	4.38	4.41

SOURCE: Abridged from D. B. Duncan, "Multiple range and multiple F tests," *Biometrics*, 11: 1–42 (1955), with the permission of the editor and the author.

Table A-3. Continued.

SIGNIFICANT STUDENTIZED RANGES FOR 5% AND 1% LEVEL NEW MULTIPLE-RANGE TEST

Error df	Protection level	\(p\) = number of means for range being tested													
		2	3	4	5	6	7	8	9	10	12	14	16	18	20
1	.05	18.0	18.0	18.0	18.0	18.0	18.0	18.0	18.0	18.0	18.0	18.0	18.0	18.0	18.0
	.01	90.0	90.0	90.0	90.0	90.0	90.0	90.0	90.0	90.0	90.0	90.0	90.0	90.0	90.0
2	.05	6.09	6.09	6.09	6.09	6.09	6.09	6.09	6.09	6.09	6.09	6.09	6.09	6.09	6.09
	.01	14.0	14.0	14.0	14.0	14.0	14.0	14.0	14.0	14.0	14.0	14.0	14.0	14.0	14.0
3	.05	4.50	4.50	4.50	4.50	4.50	4.50	4.50	4.50	4.50	4.50	4.50	4.50	4.50	4.50
	.01	8.26	8.5	8.6	8.7	8.8	8.9	8.9	9.0	9.0	9.0	9.1	9.2	9.3	9.3
4	.05	3.93	4.01	4.02	4.02	4.02	4.02	4.02	4.02	4.02	4.02	4.02	4.02	4.02	4.02
	.01	6.51	6.8	6.9	7.0	7.1	7.1	7.2	7.2	7.3	7.3	7.4	7.4	7.5	7.5
5	.05	3.64	3.74	3.79	3.83	3.83	3.83	3.83	3.83	3.83	3.83	3.83	3.83	3.83	3.83
	.01	5.70	5.96	6.11	6.18	6.26	6.33	6.40	6.44	6.5	6.6	6.6	6.7	6.7	6.8
6	.05	3.46	3.58	3.64	3.68	3.68	3.68	3.68	3.68	3.68	3.68	3.68	3.68	3.68	3.68
	.01	5.24	5.51	5.65	5.73	5.81	5.88	5.95	6.00	6.0	6.1	6.2	6.2	6.3	6.3
7	.05	3.35	3.47	3.54	3.58	3.60	3.61	3.61	3.61	3.61	3.61	3.61	3.61	3.61	3.61
	.01	4.95	5.22	5.37	5.45	5.53	5.61	5.69	5.73	5.8	5.8	5.9	5.9	6.0	6.0
8	.05	3.26	3.39	3.47	3.52	3.55	3.56	3.56	3.56	3.56	3.56	3.56	3.56	3.56	3.56
	.01	4.74	5.00	5.14	5.23	5.32	5.40	5.47	5.51	5.5	5.6	5.7	5.7	5.8	5.8
9	.05	3.20	3.34	3.41	3.47	3.50	3.52	3.52	3.52	3.52	3.52	3.52	3.52	3.52	3.52
	.01	4.60	4.86	4.99	5.08	5.17	5.25	5.32	5.36	5.4	5.5	5.5	5.6	5.7	5.7
10	.05	3.15	3.30	3.37	3.43	3.46	3.47	3.47	3.47	3.47	3.47	3.47	3.47	3.47	3.48
	.01	4.48	4.73	4.88	4.96	5.06	5.13	5.20	5.24	5.28	5.36	5.42	5.48	5.54	5.55
11	.05	3.11	3.27	3.35	3.39	3.43	3.44	3.45	3.46	3.46	3.46	3.46	3.46	3.47	3.48
	.01	4.39	4.63	4.77	4.86	4.94	5.01	5.06	5.12	5.15	5.24	5.28	5.34	5.38	5.39
12	.05	3.08	3.23	3.33	3.36	3.40	3.42	3.44	3.44	3.46	3.46	3.46	3.46	3.47	3.48
	.01	4.32	4.55	4.68	4.76	4.81	4.92	4.96	5.02	5.07	5.13	5.17	5.22	5.24	5.26
13	.05	3.06	3.21	3.30	3.35	3.38	3.41	3.42	3.44	3.45	3.45	3.46	3.46	3.47	3.47
	.01	4.26	4.48	4.62	4.69	4.74	4.84	4.88	4.94	4.98	5.04	5.08	5.13	5.14	5.15
14	.05	3.03	3.18	3.27	3.33	3.37	3.39	3.41	3.42	3.44	3.45	3.46	3.46	3.47	3.47
	.01	4.21	4.42	4.55	4.63	4.70	4.78	4.83	4.87	4.91	4.96	5.00	5.04	5.06	5.07
15	.05	3.01	3.16	3.25	3.31	3.36	3.38	3.40	3.42	3.43	3.44	3.45	3.46	3.47	3.47
	.01	4.17	4.37	4.50	4.58	4.64	4.72	4.77	4.81	4.84	4.90	4.94	4.97	4.99	5.00

Table A-4

UPPER PERCENTAGE POINTS OF THE STUDENTIZED RANGE, $q_x = \dfrac{\bar{x}_{max} - \bar{x}_{min}}{s_{\bar{x}}}$

Error df	α	p = number of									
		2	3	4	5	6	7	8	9	10	11
5	.05	3.64	4.60	5.22	5.67	6.03	6.33	6.58	6.80	6.99	7.17
	.01	5.70	6.97	7.80	8.42	8.91	9.32	9.67	9.97	10.24	10.48
6	.05	3.46	4.34	4.90	5.31	5.63	5.89	6.12	6.32	6.49	6.65
	.01	5.24	6.33	7.03	7.56	7.97	8.32	8.61	8.87	9.10	9.30
7	.05	3.34	4.16	4.68	5.06	5.36	5.61	5.82	6.00	6.16	6.30
	.01	4.95	5.92	6.54	7.01	7.37	7.68	7.94	8.17	8.37	8.55
8	.05	3.26	4.04	4.53	4.89	5.17	5.40	5.60	5.77	5.92	6.05
	.01	4.74	5.63	6.20	6.63	6.96	7.24	7.47	7.68	7.87	8.03
9	.05	3.20	3.95	4.42	4.76	5.02	5.24	5.43	5.60	5.74	5.87
	.01	4.60	5.43	5.96	6.35	6.66	6.91	7.13	7.32	7.49	7.65
10	.05	3.15	3.88	4.33	4.65	4.91	5.12	5.30	5.46	5.60	5.72
	.01	4.48	5.27	5.77	6.14	6.43	6.67	6.87	7.05	7.21	7.36
11	.05	3.11	3.82	4.26	4.57	4.82	5.03	5.20	5.35	5.49	5.61
	.01	4.39	5.14	5.62	5.97	6.25	6.48	6.67	6.84	6.99	7.13
12	.05	3.08	3.77	4.20	4.51	4.75	4.95	5.12	5.27	5.40	5.51
	.01	4.32	5.04	5.50	5.84	6.10	6.32	6.51	6.67	6.81	6.94
13	.05	3.06	3.73	4.15	4.45	4.69	4.88	5.05	5.19	5.32	5.43
	.01	4.26	4.96	5.40	5.73	5.98	6.19	6.37	6.53	6.67	6.79
14	.05	3.03	3.70	4.11	4.41	4.64	4.83	4.99	5.13	5.25	5.36
	.01	4.21	4.89	5.32	5.63	5.88	6.08	6.26	6.41	6.54	6.66
15	.05	3.01	3.67	4.08	4.37	4.60	4.78	4.94	5.08	5.20	5.31
	.01	4.17	4.83	5.25	5.56	5.80	5.99	6.16	6.31	6.44	6.55
16	.05	3.00	3.65	4.05	4.33	4.56	4.74	4.90	5.03	5.15	5.26
	.01	4.13	4.78	5.19	5.49	5.72	5.92	6.08	6.22	6.35	6.46
17	.05	2.98	3.63	4.02	4.30	4.52	4.71	4.86	4.99	5.11	5.21
	.01	4.10	4.74	5.14	5.43	5.66	5.85	6.01	6.15	6.27	6.38
18	.05	2.97	3.61	4.00	4.28	4.49	4.67	4.82	4.96	5.07	5.17
	.01	4.07	4.70	5.09	5.38	5.60	5.79	5.94	6.08	6.20	6.31
19	.05	2.96	3.59	3.98	4.25	4.47	4.65	4.79	4.92	5.04	5.14
	.01	4.05	4.67	5.05	5.33	5.55	5.73	5.89	6.02	6.14	6.25
20	.05	2.95	3.58	3.96	4.23	4.45	4.62	4.77	4.90	5.01	5.11
	.01	4.02	4.64	5.02	5.29	5.51	5.69	5.84	5.97	6.09	6.19
24	.05	2.92	3.53	3.90	4.17	4.37	4.54	4.68	4.81	4.92	5.01
	.01	3.96	4.54	4.91	5.17	5.37	5.54	5.69	5.81	5.92	6.02
30	.05	2.89	3.49	3.84	4.10	4.30	4.46	4.60	4.72	4.83	4.92
	.01	3.89	4.45	4.80	5.05	5.24	5.40	5.54	5.65	5.76	5.85
40	.05	2.86	3.44	3.79	4.04	4.23	4.39	4.52	4.63	4.74	4.82
	.01	3.82	4.37	4.70	4.93	5.11	5.27	5.39	5.50	5.60	5.69
60	.05	2.83	3.40	3.74	3.98	4.16	4.31	4.44	4.55	4.65	4.73
	.01	3.76	4.28	4.60	4.82	4.99	5.13	5.25	5.36	5.45	5.53
120	.05	2.80	3.36	3.69	3.92	4.10	4.24	4.36	4.48	4.56	4.64
	.01	3.70	4.20	4.50	4.71	4.87	5.01	5.12	5.21	5.30	5.38
∞	.05	2.77	3.31	3.63	3.86	4.03	4.17	4.29	4.39	4.47	4.55
	.01	3.64	4.12	4.40	4.60	4.76	4.88	4.99	5.08	5.16	5.23

SOURCE: This table is abridged from Table 29, *Biometrika Tables for Statisticians*, vol. 1. Cambridge University Press, 1954. It is reproduced with permission of the *Biometrika* trustees and the editors, E. S. Pearson and H. O. Hartley. The original work appeared in a paper by J. M. May, "Extended and corrected tables of the upper percentage points of the 'Studentized' range," *Biometrika*, **39**: 192–193 (1952).

Table A-4. Continued

Upper Percentage Points of the Studentized Range, $q_\alpha = \dfrac{\bar{x}_{max} - \bar{x}_{min}}{s_{\bar{x}}}$

treatment means									α	Error df
12	13	14	15	16	17	18	19	20		
7.32	7.47	7.60	7.72	7.83	7.93	8.03	8.12	8.21	.05	5
10.70	10.89	11.08	11.24	11.40	11.55	11.68	11.81	11.93	.01	
6.79	6.92	7.03	7.14	7.24	7.34	7.43	7.51	7.59	.05	6
9.49	9.65	9.81	9.95	10.08	10.21	10.32	10.43	10.54	.01	
6.43	6.55	6.66	6.76	6.85	6.94	7.02	7.09	7.17	.05	7
8.71	8.86	9.00	9.12	9.24	9.35	9.46	9.55	9.65	.01	
6.18	6.29	6.39	6.48	6.57	6.65	6.73	6.80	6.87	.05	8
8.18	8.31	8.44	8.55	8.66	8.76	8.85	8.94	9.03	.01	
5.98	6.09	6.19	6.28	6.36	6.44	6.51	6.58	6.64	.05	9
7.78	7.91	8.03	8.13	8.23	8.32	8.41	8.49	8.57	.01	
5.83	5.93	6.03	6.11	6.20	6.27	6.34	6.40	6.47	.05	10
7.48	7.60	7.71	7.81	7.91	7.99	8.07	8.15	8.22	.01	
5.71	5.81	5.90	5.99	6.06	6.14	6.20	6.26	6.33	.05	11
7.25	7.36	7.46	7.56	7.65	7.73	7.81	7.88	7.95	.01	
5.62	5.71	5.80	5.88	5.95	6.03	6.09	6.15	6.21	.05	12
7.06	7.17	7.26	7.36	7.44	7.52	7.59	7.66	7.73	.01	
5.53	5.63	5.71	5.79	5.86	5.93	6.00	6.05	6.11	.05	13
6.90	7.01	7.10	7.19	7.27	7.34	7.42	7.48	7.55	.01	
5.46	5.55	5.64	5.72	5.79	5.85	5.92	5.97	6.03	.05	14
6.77	6.87	6.96	7.05	7.12	7.20	7.27	7.33	7.39	.01	
5.40	5.49	5.58	5.65	5.72	5.79	5.85	5.90	5.96	.05	15
6.66	6.76	6.84	6.93	7.00	7.07	7.14	7.20	7.26	.01	
5.35	5.44	5.52	5.59	5.66	5.72	5.79	5.84	5.90	05	16
6.56	6.66	6.74	6.82	6.90	6.97	7.03	7.09	7.15	.01	
5.31	5.39	5.47	5.55	5.61	5.68	5.74	5.79	5.84	.05	17
6.48	6.57	6.66	6.73	6.80	6.87	6.94	7.00	7.05	.01	
5.27	5.35	5.43	5.50	5.57	5.63	5.69	5.74	5.79	.05	18
6.41	6.50	6.58	6.65	6.72	6.79	6.85	6.91	6.96	.01	
5.23	5.32	5.39	5.46	5.53	5.59	5.65	5.70	5.75	.05	19
6.34	6.43	6.51	6.58	6.65	6.72	6.78	6.84	6.89	.01	
5.20	5.28	5.36	5.43	5.49	5.55	5.61	5.66	5.71	.05	20
6.29	6.37	6.45	6.52	6.59	6.65	6.71	6.76	6.82	.01	
5.10	5.18	5.25	5.32	5.38	5.44	5.50	5.54	5.59	.05	24
6.11	6.19	6.26	6.33	6.39	6.45	6.51	6.56	6.61	.01	
5.00	5.08	5.15	5.21	5.27	5.33	5.38	5.43	5.48	.05	30
5.93	6.01	6.08	6.14	6.20	6.26	6.31	6.36	6.41	.01	
4.91	4.98	5.05	5.11	5.16	5.22	5.27	5.31	5.36	.05	40
5.77	5.84	5.90	5.96	6.02	6.07	6.12	6.17	6.21	.01	
4.81	4.88	4.94	5.00	5.06	5.11	5.16	5.20	5.24	.05	60
5.60	5.67	5.73	5.79	5.84	5.89	5.93	5.98	6.02	.01	
4.72	4.78	4.84	4.90	4.95	5.00	5.05	5.09	5.13	.05	120
5.44	5.51	5.56	5.61	5.66	5.71	5.75	5.79	5.83	.01	
4.62	4.68	4.74	4.80	4.85	4.89	4.93	4.97	5.01	.05	∞
5.29	5.35	5.40	5.45	5.49	5.54	5.57	5.61	5.65	.01	

Table A-5

SIGNIFICANT VALUES OF r AND R

Error df	P	Independent variables				Error df	P	Independent variables			
		1	2	3	4			1	2	3	4
1	.05	.997	.999	.999	.999	24	.05	.388	.470	.523	.562
	.01	1.000	1.000	1.000	1.000		.01	.496	.565	.609	.642
2	.05	.950	.975	.983	.987	25	.05	.381	.462	.514	.553
	.01	.990	.995	.997	.998		.01	.487	.555	.600	.633
3	.05	.878	.930	.950	.961	26	.05	.374	.454	.506	.545
	.01	.959	.976	.983	.987		.01	.478	.546	.590	.624
4	.05	.811	.881	.912	.930	27	.05	.367	.446	.498	.536
	.01	.917	.949	.962	.970		.01	.470	.538	.582	.615
5	.05	.754	.836	.874	.898	28	.05	.361	.439	.490	.529
	.01	.874	.917	.937	.949		.01	.463	.530	.573	.606
6	.05	.707	.795	.839	.867	29	.05	.355	.432	.482	.521
	.01	.834	.886	.911	.927		.01	.456	.522	.565	.598
7	.05	.666	.758	.807	.838	30	.05	.349	.426	.476	.514
	.01	.798	.855	.885	.904		.01	.449	.514	.558	.591
8	.05	.632	.726	.777	.811	35	.05	.325	.397	.445	.482
	.01	.765	.827	.860	.882		.01	.418	.481	.523	.556
9	.05	.602	.697	.750	.786	40	.05	.304	.373	.419	.455
	.01	.735	800	.836	.861		.01	.393	.454	.494	.526
10	.05	.576	.671	.726	.763	45	.05	.288	.353	.397	.432
	.01	.708	.776	.814	.840		.01	.372	.430	.470	.501
11	.05	.553	.648	.703	.741	50	.05	.273	.336	.379	.412
	.01	.684	.753	.793	.821		.01	.354	.410	.449	.479
12	.05	.532	.627	.683	.722	60	.05	.250	.308	.348	.380
	.01	.661	.732	.773	.802		.01	.325	.377	.414	.442
13	.05	.514	.608	.664	.703	70	.05	.232	.286	.324	.354
	.01	.641	.712	.755	.785		.01	.302	.351	.386	.413
14	.05	.497	.590	.646	.686	80	.05	.217	.269	.304	.332
	.01	.623	.694	.737	.768		.01	.283	.330	.362	.389
15	.05	.482	.574	.630	.670	90	.05	.205	.254	.288	.315
	.01	.606	.677	.721	.752		.01	.267	.312	.343	.368
16	.05	.468	.559	.615	.655	100	.05	.195	.241	.274	.300
	.01	.590	.662	.706	.738		.01	.254	.297	.327	.351
17	.05	.456	.545	.601	.641	125	.05	.174	.216	.246	.269
	.01	.575	.647	.691	.724		.01	.228	.266	.294	.316
18	.05	.444	.532	.587	.628	150	.05	.159	.198	.225	.247
	.01	.561	.633	.678	.710		.01	.208	.244	.270	.290
19	.05	.433	.520	.575	.615	200	.05	.138	.172	.196	.215
	.01	.549	.620	.665	.698		.01	.181	.212	.234	.253
20	.05	.423	.509	.563	.604	300	.05	.113	.141	.160	.176
	.01	.537	.608	.652	.685		.01	.148	.174	.192	.208
21	.05	.413	.498	.522	.592	400	.05	.098	.122	.139	.153
	.01	.526	.596	.641	.674		.01	.128	.151	.167	.180
22	.05	.404	.488	.542	.582	500	.05	.088	.109	.124	.137
	.01	.515	.585	.630	.663		.01	.115	.135	.150	.162
23	.05	.396	.479	.532	.572	1,000	.05	.062	.077	.088	.097
	.01	.505	.574	.619	.652		.01	.081	.096	.106	.115

SOURCE: Reproduced from G. W. Snedecor, *Statistical Methods*, 4th ed, The Iowa State College Press, Ames, Iowa, 1946, with permission of the author and publisher.

Table A-6. Greek Alphabet

(Letter and Name)

A	α	Alpha	H	η	Eta	N	ν	Nu	T	τ	Tau
B	β	Beta	Θ	θ	Theta	Ξ	ξ	Xi	Υ	υ	Upsilon
Γ	γ	Gamma	I	ι	Iota	O	o	Omicron	Φ	ϕ	Phi
Δ	δ	Delta	K	κ	Kappa	Π	π	Pi	X	χ	Chi
E	ϵ	Epsilon	Λ	λ	Lambda	P	ρ	Rho	Ψ	ψ	Psi
Z	ζ	Zeta	M	μ	Mu	Σ	σ	Sigma	Ω	ω	Omega

Index